Advances in Thermodynamics and Circular Thermoeconomics

SCIENCES

Energy, Field Directors – Alain Dollet, Pascal Brault

Physics of Energy and Energy Efficiency,
Subject Head – Michel Feidt

Advances in Thermodynamics and Circular Thermoeconomics

Fundamentals and Criteria

Coordinated by
Michel Feidt
Antonio Valero-Capilla

iSTE

WILEY

First published 2023 in Great Britain and the United States by ISTE Ltd and John Wiley & Sons, Inc.

ISTE Ltd
27-37 St George's Road
London SW19 4EU
UK

www.iste.co.uk

John Wiley & Sons, Inc.
111 River Street
Hoboken, NJ 07030
USA

www.wiley.com

Library of Congress Control Number: 2022949592

British Library Cataloguing-in-Publication Data
A CIP record for this book is available from the British Library
ISBN 978-1-78945-126-9

ERC code:
PE8 Products and Processes Engineering
PE8_6 Energy processes engineering
SH2 Institutions, Values, Environment and Space
SH2_8 Energy, transportation and mobility

Contents

Foreword

Michel FEIDT
LEMTA, Université de Lorraine, Vandoeuvre-les-Nancy, France

The book you have in your hands is one of the books from the "Physics of Energy and Energy Efficiency" subject in the Engineering and Systems department.

The subject of "Physics of Energy and Energy Efficiency", albeit recent, is not new. It is particularly underpinned by a thermodynamic approach, whatever the scale.

The selected aspect will be phenomenological and characterized explicitly in order to emphasize the key concept of "efficiency", essential for any system or process.

The characterization chosen for the development of this subject has been arranged into four successive books, each strongly correlated with each other, and also with other series within the department:

– *Fundamental Physics of Energy*;

– *Thermodynamics of Heat Engines*;

– *Heat Engines with Inverse Cycles*;

– *Efficiency in Practice*.

Advances in Thermodynamics and Circular Thermoeconomics,
coordinated by Michel FEIDT and Antonio VALERO-CAPILLA.

I would like to thank ISTE, the various coordinators, and the authors for their contributions and effective actions, despite the very particular conditions of the moment. We are awaiting and would like to encourage comments, suggestions and questions from readers.

Introduction

Michel FEIDT[1] **and Antonio VALERO-CAPILLA**[2]
[1] *LEMTA, Université de Lorraine, Vandoeuvre-les-Nancy, France*
[2] *CIRCE Institute, University of Zaragoza, Spain*

This book on energy physics and energy efficiency discusses two essential components of energy physics: the fundamentals and the criteria.

The fundamentals essential to energy physics are fairly recent (about two centuries old) under the impetus of S. Carnot.

This introduction offers a selective review of previous work in French-language journals. The purpose of this introduction is to make a general point on the topic of thermomechanical systems and processes, and more particularly the so-called heat engines, which are what sparked the industrial revolution.

Among these engines are:

– Brayton-Joule engines, commonly referred to as gas or combustion turbines (Remy et al. 1994; Feidt 1996; Molière 2000; Chen and Su 2005; Wang et al. 2005; Ust et al. 2006; Tyagi et al. 2006);

– internal combustion engines, including diesel engines (or diesel cycle) (Chesse et al. 1998; Asad and Zheng 2008; Xia et al. 2012; Czarneski et al. 2019);

– other engines, less widespread, namely:

  - Braysson heat engines (Zhou et al. 2004),

Advances in Thermodynamics and Circular Thermoeconomics,
coordinated by Michel FEIDT and Antonio VALERO-CAPILLA.

- air motors and standard Miller cycle engines (Lin and Hou 2008),

- an engine with three thermodynamic cycles (Zhao and Chen 2007),

- Stirling engines with many variants whose interest is growing (Popescu et al. 1996). The engines are particularly examined in Desmet (2023), another volume in this series.

There are also other well-studied alternative systems, mainly the fuel cell (Loizeau 1995), but also more traditional configurations such as the steam engines of thermal power plants (Radcenco et al. 2007; Erdem et al. 2009). These last two articles focus on the technical and technological part, a current practical extension of the more fundamental Carnot studies.

We also include articles dealing with reverse cycle engines, given their importance in practice, but whose studies remain less numerous in the *International Journal of Thermal Sciences* (Wu et al. 2011). These machines are examined in detail in Bonjour (2023), another volume in this series, and in Chapter 2 of this volume in a more succinct form.

More innovative or more conceptual and fundamental approaches should be added to the previous approaches, including the two articles proposed by Barrère (1980, 1983). We had the chance to have a discussion with the latter author, reinforcing our interest in the work on interdependence of heat transfers and the conversion of energies in systems and processes, from a thermomechanical stance.

This theme has since been supplemented by:

– stability analyses (Nie et al. 2008);

– exergy analyses (Fratzscher 1997);

– entropy analyses of irreversibilities (Grazzini and Gori 1987; Lucia and Grazzini 1997).

Finally, the economic aspects are also a long-standing concern and are already linked to respect for the environment (Barrère 1983; Goffause 1994). These aspects have been masterfully developed by our Spanish colleagues in Chapter 5.

In this volume, the more fundamental aspects related to the power and efficiencies associated with this same optimized power will be covered. The challenge is to propose an engine quality criterion closer to its functional reality, as proposed in previous studies (Barrère 1980; Roche 1983; Barrère 1995; Stitou and Feidt 2005; Arias 2018) with regard to the well-known Carnot efficiency limit.

This book is based on the previously cited models (Carnot, Chambadal) to show the progression of knowledge on the existence of successive energy, power, and then efficiency optima associated with these successive optima, as well as with the production of entropy or anergy.

To do this, the following plan is adopted. Chapter 1 will develop the historical basis of Carnot models showing, if necessary, that these last models have deeper roots in the history of thermodynamics. Chapter 3 will address the optimization of thermomechanical engines by insisting on the methodology of the studies:

– different choices for the objective function;

– different choices for constraints;

– parametric sensitivity studies: distinction between variables and parameters, whose connection to boundary conditions cannot be ignored. Important consequences ensue, some of which will be reported.

A few results will illustrate the progressive (graded) approach used, leading to a sequential optimization whose outcome provides the desired optimum, corresponding (in the general irreversible case) to the Carnot efficiency, which constitutes the unattainable reversible limit.

Chapters 3 and 4 will consist of a discussion-synthesis of results unifying quite a number of works in the literature, including those published in the Revue Générale de Thermique, which then became the *International Journal of Thermal Sciences* (Feidt 1987; Feidt and Costea 2018).

In summary, Chapters 1 and 2 revisit the thermostatic cycles of double-function heat pumps by restoring the basic definitions of the efficiencies of engine or receiver systems (heat pump, refrigerator). The focus is on the graphical representation in the form of diagrams (Q-T, Borel-Favrat, Raveau, ternary diagram). The general illustration relates to thermal systems of the Carnot-Stirling-Ericsson type.

The previous graphical approach is discussed in a more analytical way in Chapters 3 and 4. These chapters detail the evolution of knowledge of equilibrium thermodynamics towards irreversible thermodynamics by illustrating the Carnot engine, its efficiency and power.

Using the Carnot engine as an example, Chapters 1 and 2 show the consubstantial links between the production of entropies and efficiency leading to:

– trade-off between energy transfer and conversion;

– the existence of a cycle period (frequency), leading to maximum engine power (new paradigm).

All this is then developed in a more complete and particular way, with Chapter 3 being dedicated to Finite Time Thermodynamics (FTT).

Chapter 4 is an overview of what we call finite physical dimensions optimal thermodynamics.

Chapter 5 describes, for the first time, the theory of "circular thermoeconomics" to assess the physical costs of recycling waste in increasingly complex industrial processes. It is demonstrated that thermoeconomics encompasses the physical principles of "circular economy" (rigorously "spiral economy") and "industrial symbiosis", in addition to Second Law costing and diagnosis complex energy systems.

Every time we produce, we produce waste. So, in addition to the cost of production, there is the cost of waste treatment. Thermoeconomics is the physical theory that allows us to analyze the process of cost formation. Cost is an economic concept, whereas irreversibility is genuinely thermodynamic. Hence the deeper message of thermoeconomics is that any conceivable activity needs to destroy exergy to perform it. From thermodynamics we know how to measure the cumulative destruction of exergy needed to carry out any process. This is done with the concept of generated entropy, S_g, which by the Gouy-Stodola Law is a measure of irreversibility, $I = T_0S_g$. In other words, the physical cost of producing something can be measured as the sum of the irreversibilities that have occurred throughout the formation of a product.

From this point of view, what does thermodynamics say about waste? Curiously enough, this problem has not been addressed in the literature. The answer to the question is as simple as this: if it is a physical consequence of production, it can and should be considered and measured as another irreversibility. In other words, the cost of producing waste is, no more and no less, another irreversibility, but this time one that is external to the production process. Or, in other words, producing something generates internal and external irreversibilities. We call the latter waste.

Accordingly, waste would cease to be waste if all its exergy, i.e., its capacity to do/harm something, vanished. Or, in other words, it would reach equilibrium with its environment (if its environment were an infinite reservoir, which in reality it is not).

Regrettably, in real life, we behave in a vicious way towards the planet. We outsource our waste in the hope that no one will pay for it. The responsible way to produce should be to internalize waste, i.e., to extinguish its exergy, and this process is just another cost (irreversibility) of production.

Thus, for the first time, the concept of circular thermoeconomics appears in the scientific literature. As long as there is exergy, it must be harnessed rather than externalized. To recover it, production loops must be closed.

Efficiency, rigorously based on the Second Law, can no longer be defined as the simple quotient between the exergy of the product divided by the exergy of the consumed resources. Waste must be taken into account. If the exergy extinction requires more exergy from the resources, it will have to be accounted for in the debt. But if this residual exergy is used to increase production, we will have achieved an overall increase in production efficiency. Reality rides between both situations, so we consider an "exergy pinch point" as a limit to our capacity to use this exergy until it is extinguished. This limit must be reasonable and reasoned to take advantage of the residual exergy.

Besides these reflections, this chapter outlines new ideas such as:

– A brief history of exergy cost theory as a basis of thermoeconomics.

– A reformulation of the Exergy Cost Theory (ECT) and its FPR rules about resources cost, cost balance, co-products and unspent fuel cost allocation and waste cost internalization, as is described in Torres and Valero (2021).

– A reformulation of the structural theory as a general cost theory, and its characteristic equations, showing that the ECT is a particular case of structural theory, where its characteristic equations are the definition of process efficiency. This theory can be used for cost allocation in cases of industrial symbiosis, process integration and waste recycling.

– The proposed methodology allows for a decomposition of production costs between costs due to internal irreversibilities, which can be reduced with improvements in design, operation and maintenance, and those due to waste, which can be reduced with process integration and recycling. It is noted that there is a pinch point beyond which it does not make sense to recycle. The *cost-irreversibility* formula presented here is the basis of the connection between thermodynamics and circular economy.

Chapter 6 presents a new thermodynamic property, “relative free energy”, which allows the construction of a new mathematical theory of thermodynamic costs to diagnose malfunctions in thermal systems. This new function plus the “deterioration temperature” parameter allows for an exact relationship between the amount of resources used and the increase in entropy generation caused by the deterioration path of the component.

This chapter goes beyond the conventional thermoeconomic diagnosis of industrial energy systems. Throughout history, the emergence of new basic properties has been a far-reaching scientific event that has opened up new avenues and applications of thermodynamics. In the same way that there is a fundamental equation of the behavior of all the equilibrium states of a system, of the type $U = U(S, V, N_i)$, we postulate the existence of a characteristic equation that describes the behavior of a flow that carries out a useful process in an engine that has deteriorated due to some cause “r”. Since the deterioration of the engine may be due to various causes, the working substance evolving inside the engine will undergo variations in enthalpy and entropy different from the design ones.

We name deterioration trajectory function $h_r = h_r(s_r)$ to the set of pairs (h_r, s_r) of the flowing substance in the deteriorated engine due to cause r. Now, using the Legendre transform this path can also be described by the intersection of the tangent line of the deterioration path r with the h-axis and its slope is T_d. In other words, the equation:

$$\ell = h - T_d s$$

contains the same information as the trajectory $h_r = h_r(s_r)$. This gives rise to the new property, the *relative free energy* together with its conjugate, the *deterioration temperature* of the substance flowing inside the engine.

We call this new function *relative free energy* (RFE) because it has energy dimensions, it is the enthalpy part of the flow that remains available to perform its task in the engine, and it is relative because it is associated to a cause, r, of engine deterioration. Note that the *deterioration temperature* has dimensions of temperature but in no way can be measured as a conventional temperature. Also note that the higher the T_d, the lower the RFE is.

Remarkably, RFE has some interesting properties in the diagnosis of energy systems. Thus, if the concept of equipment malfunction is due to several causes, each of them will have its deterioration path and deterioration temperature. We found that the amount of fluid that must compensate for the engine deterioration depends on the RFE and the deterioration temperature and it does not matter which level of aggregation the analyst chooses.

This allows us to rethink the concept of exergy cost and second law efficiency. Besides that, as the component's deterioration does not depend on the analyst's decision on the chosen reference state, the deterioration temperature no longer depends on the chosen reference temperature for exergy calculations, and neither does the new definition of efficiency using the RFE. This fact does not isolate the plant from environmental conditions, but eliminates the arbitrary selection of any reference state. This idea clarifies the role of the environmental conditions on the plant and perhaps, it will be more important to use the RFE instead of the exergy in plant diagnosis.

Our warmest thanks go to the ISTE team and the authors for their contributions. The international bibliographical references (Feidt 1987; Feidt and Costea 2018) are a deepening and an opening to the diligence of the reader.

Michel FEIDT and Antonio VALERO-CAPILLA
November 2022

References

Andresen, B. (1983). Finite-Time Thermodynamics. Physics Laboratory II, University of Copenhagen, Copenhagen.

Arias, F. (2018). On osmotic heat engines driven by thermal precipitation-dissolution of saturated aqueous solution. *Int. J. Therm. Sci.*, 133, 151–161.

Asad, U. and Zheng, M. (2008). Fast heat release characterization of a diesel engine. *Int. J. Therm. Sci.*, 47, 1688–1700.

Barrère, M. (1980). Le rôle du temps dans l'optimisation des cycles thermodynamiques. *Rev. Gen. Therm. Fr.*, 228, 995–1006.

Barrère, M. (1983). La thermoéconomie. *Rev. Gen. Therm. Fr.*, 225, 243–262.

Barrère, M. (1995). Démarche unitaire de la thermodynamique dans l'étude des systèmes complexes. *Rech. Aerospatiale*, 5, 353–362.

Blaise, M. (2016). Contribution au développement des moteurs à apport de chaleur externe utilisés pour la conversion des rejets thermiques industriels en électricité. Modélisation et optimisation thermodynamique d'un nouveau convertisseur : Turbosol. PhD Thesis, Université de Lorraine.

Bonjour, J. (2023). *Refrigerators, Heat Pumps and Reverse Cycle Engines: Principles, State of the Art and Trends.* ISTE Ltd, London, and John Wiley & Sons, New York.

Chambadal, P. (1957). *Les centrales nucléaires*. Armand Colin, Paris.

Chen, C.-K. and Su, Y.-F. (2005). Exergetic efficiency optimization for an irreversible Brayton refrigeration cycle. *Int. J. Therm. Sci.*, 44, 303–310.

Chesse, P., Hetet, J.F., Tauzia, X., Frayret, J.P. (1998). Influence of the engine/compressor duct on the operating range of a high output diesel engine with a two-stage turbo charging system. *Rev. Gen. Therm.*, 37, 801–812.

Curzon, F.L. and Ahlborn, B. (1975). Efficiency of a Carnot engine at maximum power output. *Am. J. Phys.*, 43, 22–24.

Czarneski, F.E., Och, S.H., Moura, L.M., Domingues, E. (2019). Experimental model for prediction of tip temperature of diesel injector nozzle in dual-fuel engines. *Int. J. Therm. Sci.*, 138, 596–604.

Desmet, B. (2023). *Thermodynamics of Heat Engines*. ISTE Ltd, London, and John Wiley & Sons, New York.

Erdem, H.H., Akkaya, A.V., Cetin, B., Dagdas, A., Sevilgen, S.H., Sahin, B., Teke, I., Gungor, C., Atas, S. (2009). Comparative energetic and exergetic performance analysis for coal-fired thermal power plants in Turkey. *Int. J. Therm. Sci.*, 48, 2179–2186.

Feidt, M. (1987). *Thermodynamique et optimisation énergétique des systèmes et procédés*, 1st edition. Lavoisier TEC & DOC, Paris[1].

1. See also the 1996 and 2016 editions.

Feidt, M. (1996). Optimisation d'un cycle de Brayton moteur en contact avec des capacités thermiques finies. *Rev. Gen. Therm.*, 35, 662–666.

Feidt, M. (2001). Reconsideration of efficiency of processes and systems from a non-equilibrium point of view. *Int. J. Energy Environ. Econ.*, 11, 31–49.

Feidt, M. (2012). Thermodynamics of energy systems and processes: A review and perspectives. *J. Appl. Fluid Mech.*, 5, 85–98.

Feidt, M. (2013). *Thermodynamique optimale en dimensions physiques finies.* Hermes, Paris.

Feidt, M. and Costea, M. (2018). From finite time to finite physical dimensions thermodynamics: The Carnot engine and Onsager's relations revisited. *J. Non-Equil. Thermodyn.*, 43, 151–161.

Feidt, M. and Costea, M. (2019a). A new perspective on Carnot and Curzon-Ahlborn models of thermomechanical engine by transfer and production of entropy. *15th Joint European Thermodynamic Conference, JETC 2019*, May 22–24, Barcelona.

Feidt, M. and Costea, M. (2019b). Progress in Carnot and Chambadal modeling of thermomechanical engine by considering entropy production and heat transfer entropy. *Entropy*, 21, 1232.

Feidt, M. and Feidt, R. (In press). Du rôle fondamental de l'entropie de transfert de chaleur et de la production d'entropie dans la description et l'optimisation des moteurs thermodynamiques : le cas emblématique de la machine de Carnot. *C. R. Acad. Sci.*

Feidt, M., Costea, M., Petrescu, S., Stanciu, C. (2016). Nonlinear thermodynamic analysis and optimization of a Carnot engine cycle. *Entropy*, 18, 243.

Feidt, M., Costea, M., Stanciu, C., Stanciu, D. (2019). Entropic model and optimization of a refrigeration machine. *IOP Conference Series: Materials Science and Engineering*, 595.

Fratzscher, E. (1997). Exergy and possible applications. *Rev. Gen. Therm.*, 36, 690–696.

Goffause, P. (1994). How to finance the energy future of greater Europe and respect the environment. *Rev. Gen. Therm.*, 33, 214–216.

Grazzini, G. and Gori, F. (1987). Influence of thermal irreversibilities on work producing systems. *Rev. Gen. Therm.*, 312, 637–639.

Ibrahim, O.M., Klein, S., Mitchell, J.W. (1991). Optimum heat power cycles for specified boundary conditions. *J. Eng. Gas Turb. Power*, 113, 514–521.

Lin, J.C. and Hou, S.S. (2008). Performance analysis of an air-standard Miller cycle with considerations of heat loss as a percentage of fuel's energy, friction and variable specific heats of working fluid. *Int. J. Therm. Sci.*, 47, 182–191.

Loizeau, S. (1995). Thermodynamic approach to the fuel-cell considered as an open system. *Rev. Gen. Therm.*, 34, 426–436.

Lucia, U. and Grazzini, G. (1997). Global analysis of dissipation due to irreversibility. *Rev. Gen. Therm.*, 36, 605–609.

Molière, M. (2000). Stationary gas turbines and primary energies: A review of fuel influence on energy and combustion performances. *Int. J. Therm. Sci.*, 39, 141–172.

Moutier, J. (1872). *Eléments de thermodynamique*. Gauthier-Villars, Paris.

Nie, W., He, J., Deng, X. (2008). Local stability of an irreversible Carnot heat engine. *Int. J. Therm. Sci.*, 47, 663–640.

Novikov, I. (1957). The efficiency of atomic power stations. *At. Energy*, 3, 409–412.

Petrescu, S., Costea, M., Feidt, M., Ganea, I., Boriaru, N. (2015). *Advanced Thermodynamics of Irreversible Processes with Finite Speed and Finite Dimensions*. AGIR, Bucharest.

Popescu, G., Radcenco, V., Costea, M., Feidt, M. (1996). Optimisation thermodynamique en temps finie du moteur de Stirling endo- et exo-réversible. *Rev. Gen. Therm.*, 35, 656–661.

Radcenco, V., Vasilescu, E.E., Popescu, G., Apostol, V. (2007). New approach to thermal power plants operation regimes maximum power versus maximum efficiency. *Int. J. Therm. Sci.*, 46, 1259–1266.

Remy, P., Boissenin, Y., Molière, M. (1994). Gas turbine and cogeneration installations of complex construction – The UEM power-plant case. *Rev. Gen. Therm.*, 33, 717–725.

Roche, M. (1983). Récupération de chaleur, le cycle thermodynamique idéal. *Rev. Gen. Therm.* Fr., 260–261, 561–571.

Ser, L. (1888). *Traité de physique industrielle : production et utilisation de la chaleur (Machines à air chaud)*. G. Masson, Paris.

Stitou, D. and Feidt, M. (2005). Nouveaux critères pour l'optimisation et la caractérisation des procédés thermiques de conversion énergétique. *Int. J. Therm. Sci.*, 44, 1142–1153.

Torres, C. and Valero, A. (2021). The exergy cost theory revisited. *Energies*, 14, 1594.

Tyagi, S.K., Chen, G.M., Wang, Q., Kaushik, S.C. (2006). Thermodynamic analysis and parametric study of an irreversible regenerative-intercooled-reheat Brayton cycle. *Int. J. Therm. Sci.*, 45, 829–840.

Ust, Y., Sahin, B., Kodal, A. (2006). Performance analysis of an irreversible Brayton heat engine based on ecological coefficient of performance criterion. *Int. J. Therm. Sci.*, 45, 94–101.

Vaudrey, A., Lanzetta, F., Feidt, M. (2014). H. B. Reitlinger and the origin of the efficiency at maximum power for heat engines. *J. Non-Equil. Thermodyn.*, 39, 199–203.

Wang, W., Chen, L., Sun, F., Wu, C. (2005). Power optimization of an endoreversible closed intercooled regenerated Brayton cycle. *Int. J. Therm. Sci.*, 44, 89–94.

Wu, W.D., Zhang, H., Men, C.L. (2011). Performance of a modified zeolite 13Xwater adsorptive cooling module powered by exhaust waste heat. *Int. J. Therm. Sci.*, 50, 2042–2049.

Xia, S., Chen, L., Sun, F. (2012). Engine performance improved by controlling piston motion: Linear phenomenological law system diesel engine. *Int. J. Therm. Sci.*, 51, 163–174.

Zhao, Y. and Chen, J. (2007). An irreversible heat engine model including three typical thermodynamic cycles and their optimum performance analysis. *Int. J. Therm. Sci.*, 46, 605–613.

Zhou, Y., Tyagi, S.K., Chen, J. (2004). Performance analysis and optimum criteria of an irreversible Braysson heat engine. *Int. J. Therm. Sci.*, 43, 1101–1106.

1

From Equilibrium Thermodynamics to Irreversible Thermodynamics

Michel FEIDT
LEMTA, Université de Lorraine, Vandoeuvre-les-Nancy, France

1.1. Recent emergence of thermodynamics: from heat to engines

1.1.1. *Heat and temperature*

The mastery of energy is a story as old as humanity. Very early on, humans tried to master fire (heat) for their basic needs (heating, cooking, security and light). Maintaining a warm atmosphere calls on the notion of heat and, from a practical point of view, combustion and fuel.

However, it took a very long time to go from use to concepts. The first form of energy thus appears as *thermal energy*. The associated formalization is quantitative: Q represents a quantity of heat. This heat is most often sensible heat. Initial modeling therefore shows, for an elementary transformation of a system transitioning from an initial state to a final state (in a time interval dt), a variation δQ (differential form) where:

$$\delta Q = C \cdot \mathrm{d}T \qquad [1.1]$$

The quantity of heat exchanged (received or transferred) is proportional to the variation in temperature of the material of the system (mass M)

Advances in Thermodynamics and Circular Thermoeconomics,
coordinated by Michel FEIDT and Antonio VALERO-CAPILLA.

through its calorific capacity. This calorimetric equation can be specified in the more generic form:

$$C = M \cdot c \quad [1.2]$$

where c represents a specific heat capacity.

This quantity can depend on the temperature, but in particular on the way in which the heat is transmitted to the system: most often, in an isobaric (c_p) or isochoric (c_V) way.

Thermodynamic science therefore arose from calorimetry and thermometry in the 19th century.

For more generality, physicists (Clausius) introduced the notion of entropy, S, as a conjugate variable of temperature (S, extensity; T, intensity), for the particular form of energy that is heat. The temperature thus appears as a thermal potential.

After various attempts and the development of temperature measurements, current physics is based on absolute temperature (which is expressed in Kelvin), with this temperature being linked to the current temperature in °C (Celsius) by the relationship:

$$T_K = T_0 + t_{°C} \quad [1.3]$$

with $T_0 = 273.15$ K.

We will come back to these two quantities later in the book, given their fundamental importance.

1.1.2. *Matter and chemical reactions*

The combustion used very early on by humanity allows us to see the transformation of matter: wood, a noble form of matter, becomes ash and smoke.

We had to wait for the work of Lavoisier to understand that on a human scale, matter is conserved. Hence, the use of chemical equations, for example:

$$H_2 + \frac{1}{2}O_2 \rightarrow H_2O \qquad [1.4]$$

A pure substance can be found in three forms: solids (S), liquids (L) and gases (G). Everyone appreciates this fact, looking at water, an essential element for life, which is most often found on Earth in the liquid form, but also as vapor and sometimes solid (ice or snow). The transition between these various phases can only take place in the presence of a heat transfer different from the transfer of sensible heat. This is a latent heat L, which for a pure substance is a characteristic of temperature or pressure.

In the context of this book, we will not go any deeper and the interested reader can find additional information in other specialized works (Feidt 2016).

We will conclude this section with the conservation of matter, which was later extended to the conservation of electric charges.

1.1.3. *Mechanical energy*

Mechanics very often use machines involving the flow of liquid or gaseous fluids. Viewing the conservation of matter as a continuous medium then leads to the equation:

$$div\, \rho\vec{v} + \frac{\partial\rho}{\partial t} = 0 \qquad [1.5]$$

In steady state, this relation simplifies to:

$$div\, \rho\vec{v} = 0 \qquad [1.6]$$

and for a liquid which can be considered quasi-incompressible (ρ, constant density):

$$div\, \vec{v} = 0 \qquad [1.6']$$

where $\vec{v}$ represents the local flow velocity vector.

A similar reasoning can be carried out for momentum; this will not be discussed here. We will simply note that for ideal fluids (negligible viscosity effects), whose density is considered constant, we arrive at the Bernoulli equation, which introduces the notion of total pressure by the sum of a driving pressure ($p + \rho gz$) and a dynamic pressure ($\frac{1}{2}\rho v^2$). This same equation can be expressed in a reduced energy form $e = E/\rho$ (J/kg):

$$e = \frac{p}{\rho} + gz + \frac{v^2}{2} \quad [1.7]$$

This is a form of conservation of energy, which we will return to later.

Indeed, for real fluids, one cannot ignore the influence of viscous friction (or turbulence). These phenomena induce the *degradation of mechanical energy* into heat. As a result, the total mechanical energy is no longer conserved. There appears to be a loss of charge ΔP (or pressure) along the flow. A distinction is often made between two types of pressure loss: *regular pressure losses* and *singular pressure losses*. It is difficult to have precise models of these energy degradation phenomena. Most often the approaches remain of the empirical or semi-empirical type (see, for example, Idel'cik (1986)).

1.1.4. *Heat–work equivalence*

The previous section introduced conservation laws (matter, electric charge, mechanical quantities). However, energy degradation phenomena also appeared. The diagram below shows degradation of mechanical energy by solid–solid friction.

Fluid and solid mechanical friction gave rise to an important experiment, the Joule experiment (Figure 1.1).

This experiment made it possible to highlight the equivalence of heat and mechanical work: the two thermal and mechanical energies are expressed in Joules, even if their forms are fundamentally different.

In other words, the energy is quantitatively conserved, whatever its form. But on the other hand, the quality of the energy varies; it is degraded during the evolutions. This important observation will be discussed later in the book.

In the immediate sections, all forms of energy are deemed "noble", except heat which cannot be completely converted into noble energies.

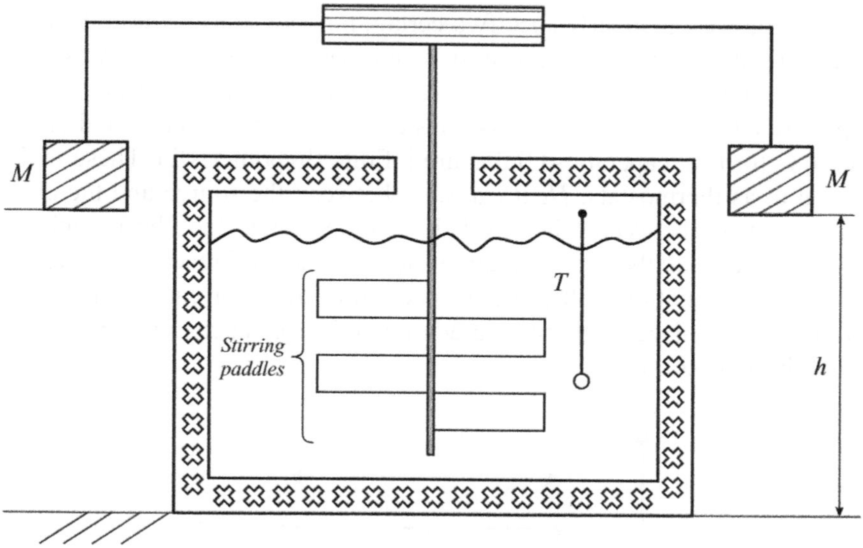

Figure 1.1. *Joule experiment (Feidt 2013)*

1.2. From engines to concepts (work by Carnot)

1.2.1. *The steam engine and other engines*

The real birth of thermodynamics occurred as a result of previous remarks and from the revelatory work of Carnot (1824), who is considered the father of current thermodynamics.

Carnot, because of his father, was sensitive to the use of mechanical energy. He was interested in the recent development of thermomechanical engines aimed at increasing the power available in mechanical form tenfold for the needs of a nascent industry (from human energy, to using horses, then to horsepower).

1.2.2. *The Carnot cycle*

Carnot had an intuition about the importance of cyclic transformations, and therefore of the notion of the thermodynamic cycle and the cycle that now bears his name. This cycle is represented in an entropy diagram (T, S) in the form indicated in Figure 1.2.

It is necessary to clearly specify a certain number of hypotheses to which we can return later. Indeed, Carnot's approach is limiting because it is located within the restricted framework of equilibrium thermodynamics (hypothesis 1). Furthermore, it assumes a supposed heat source and sink of reservoirs (or heat capacities C) that are infinite (hypothesis 2). These will be referred to as thermostats. *Heat transfers* between the source and the cycled medium, then between the cycled medium and the heat sink are assumed to be ideal: they do not require a temperature difference (hypothesis 3). Finally, the circulation of the cycled medium (a liquid or gaseous fluid, or a mixture) takes place without degradation of mechanical energy, that is, reversibly (hypothesis 4).

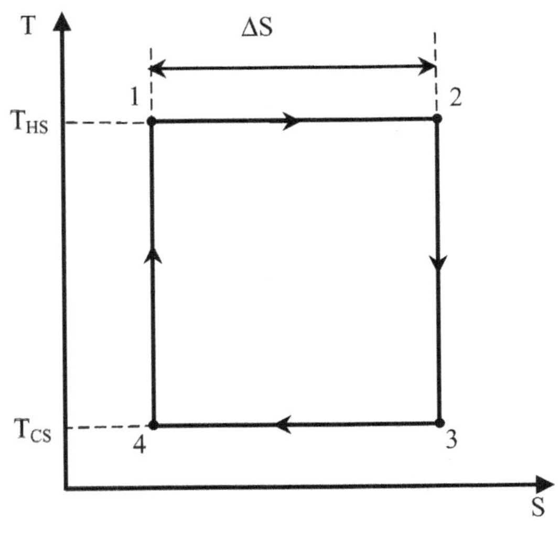

Figure 1.2. *Carnot engine cycle according to equilibrium thermodynamics*

Under these conditions, the thermodynamic system consists of two thermostats and *the thermomechanical converter*, these three elements are assumed to be insulated from the outside (no heat losses, hypothesis 5).

Then the engine operates according to the engine cycle 1-2-3-4.

The 1-2 transformation is an isothermal transformation at T_{HS}, temperature of the hot thermostat. In this transformation, the cycled medium receives the amount of heat Q_{HS} from the source.

The 2-3 transformation is an adiabatic transformation (without heat exchange of the medium) and reversible, that is to say isentropic (S = constant). The transition from temperature T_{HS} to temperature T_{CS} corresponds to a pressure loss.

The 3-4 transformation is an isothermal transformation at T_{CS}, reciprocal of the 1-2 transformation. It is accompanied by the transfer of an amount of heat Q_{CS} from the medium to the cold sink.

The 4-1 transformation and in turn the reciprocal of the 2-3 transformation (transition from T_{CS} to T_{HS}) corresponds to an increase in pressure.

Here, note that the points 1, 2, 3, 4 correspond to ideal instantaneous switching (hypothesis 6) between the successive transformations.

In summary, the proposed system is a double-function heat pump system (with source and sink thermostats) involving *heat transfers* at the hot and cold end. The passage through the intermediate thermomechanical converter allows the production of useful mechanical energy W deduced from the energy balance over the cycle (conservation of energy):

$$W = Q_{HS} - Q_{CS} \qquad [1.8]$$

All quantities will be > 0 (standard) in accordance with section 1.1.1, it becomes:

$$Q_{HS} = T_{HS} \cdot \Delta S; \quad Q_{CS} = T_{CS} \cdot \Delta S \qquad [1.9 \text{ and } 1.10]$$

and so

$$W = (T_{HS} - T_{CS}) \cdot \Delta S \qquad [1.11]$$

Consequently, it is recognized that a single-heat source engine cannot operate ($T_{HS} = T_{CS} \Rightarrow W = 0$), as heat transfers are necessary ($\Delta S \neq 0$).

All of these quantities appear in the diagram in geometric form.

1.2.3. *Carnot efficiency*

Carnot had the good idea of extending the notion of mechanical efficiency to thermomechanical engines. We are therefore taking advantage of this section to give a definition of performance to which we will return in the context of the concept of efficiency (see the fourth book of the series). This definition, although general, will be particularized here, within the framework of the first law of thermodynamics: η_I represents the *engine efficiency in the sense of the first law*. It is defined more generally as the ratio of a useful effect (W, for the engine), to the effort (expense) granted here Q_{HS}, whereby:

$$\eta_I = \frac{W}{Q_{HS}} \qquad [1.12]$$

Using relation [1.8], it becomes:

$$\eta_I = 1 - \frac{Q_{CS}}{Q_{HS}} \qquad [1.13]$$

Using the relations [1.9] and [1.10], we get:

$$\eta_I = 1 - \frac{T_{CS}}{T_{HS}} \qquad [1.14]$$

In conclusion, within the framework of the hypotheses that we have discussed (i.e. equilibrium thermodynamics), the efficiency in the sense of the first law depends only on the temperatures of the sink and the source. This efficiency is always less than 1; the conversion of heat into (noble) mechanical energy can never be complete, unless the cold sink has a temperature of 0 K, which we know is unattainable according to the third law (Feidt 1987).

Moreover, equilibrium thermodynamics imposes that all the transformations be quasi-static (their durations, Δt_i must tend to infinity). Remember that Carnot spoke of motive power. The consequences of this quasi-staticity will be discussed in the following.

1.2.4. *Engine power*

If Δt is the cycle time, the combination of Δt with the relation [1.8] gives average engine power $\dot{W}$:

$$\dot{W} = \frac{W}{\Delta t} = \frac{Q_{HS}}{\Delta t} - \frac{Q_{CS}}{\Delta t} \quad [1.15]$$

where $\dot{W}$ is flow of mechanical energy (or power), $\dot{Q}_{HS}$ is heat flux provided by the source and $\dot{Q}_{CS}$ is heat flow transferred to the cold sink.

Here, the essential difference between the flow of G, denoted $\dot{G}$, which for energies is measured in W, and the variation in time of G, $\frac{dG}{dt}$, which corresponds to a capacitive effect (storage or release depending on the sign). It corresponds to transient behaviors (any variations over time).

On the contrary, for a system in steady state, the fluxes are by definition constant whatever the time, but they vary in space.

The consequence of the previous points is that the equilibrium thermodynamics, if it accounts for the maximum efficiency [1.14] possible within the framework of its assumptions, leads to zero powers of the associated engines per [1.15].

As a result, there is no Carnot engine providing such a power, neither in theory nor in practice. Nevertheless, it is possible to calculate the efficiency for a real engine in the sense of the first law η_{IR}, from the heat fluxes transferred to the hot and cold end by:

$$\eta_{IR}(\dot{W}) = 1 - \frac{\dot{Q}_{CS}}{\dot{Q}_{HS}} \quad [1.16]$$

This efficiency is lower than the Carnot efficiency. In the next section, the shortcomings of equilibrium thermodynamics and how to overcome them will be discussed in detail. The proposals are focused on the concept of energy in the case of a real engine, but power modeling is also introduced in the simplest possible way.

1.3. From thermostatics to thermodynamics

This section discusses the main part of thermostatics (or equilibrium thermodynamics, in section 1.3.1) supplemented by transfers of matter and energy (in section 1.3.2). Indeed, these notions are necessary to develop the thermodynamics of systems and processes.

There is a differentiation between transfers and conversions (in section 1.3.3.1) for the different forms of energy. Highlighting this duality is fundamental.

The section will conclude with a generalization of the concept of cycles and efficiency (in section 1.3.4).

1.3.1. *The basics of thermodynamics*

The first essential notion is that of a system.

1.3.1.1. *Thermodynamic system*

The studied system is defined in space by its control volume V_C (see Figure 1.3) at time t.

This system is delimited from the surroundings by boundaries and interacts with the universe according to properties, which must be carefully specified. For this, at given t, it is necessary to know the local properties of any point P inside the elementary volume dV_C (local approach), as well as the properties of any boundary element dA_C of the overall boundary A_C.

These conditions are difficult to fully satisfy. Also, they are often approximated by averaging the properties. Simple cases are associated with the conditions of isotropy and homogeneity.

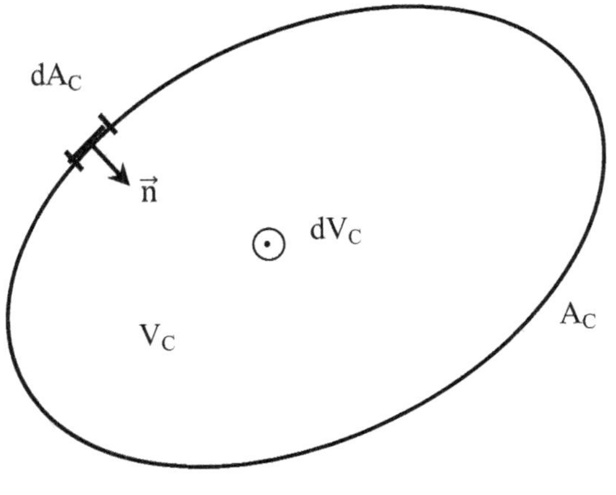

Figure 1.3. *Representation of a thermodynamic system (at time t)*

Among the simple cases are:

– *isolated systems*: no matter or energy is exchanged with the exterior;

– *closed systems*: can exchange all forms of energy with the exterior, but not matter. The exchange often occurs with the normal $\vec{n}$ directed toward the inside of the control volume V_C;

– *open systems*: exchange of matter and energy with the exterior. This scenario is the most common case in practice.

1.3.1.2. *System state*

This state is characterized by various properties:

– *scalar* (ρ, density; P, pressure; T, temperature);

– *vector* (point velocity, $\overrightarrow{V(M)}$; point acceleration, $\overrightarrow{\gamma(M)}$; electric field, $\vec{E}$; magnetic field, $\vec{H}$);

– *tensor* (stress tensor, $\overline{\overline{\sigma(M)}}$; strain tensor, $\overline{\overline{\epsilon(M)}}$).

It appears that the most general state is a *local function* at point M. Most often, for simplification reasons, this is discussed from a *general* point of view (as averages and not at one point).

1.3.1.3. *State variables*

This concept is essential in thermodynamics. Thus, for a closed thermodynamic system, the internal energy U of a system can be considered as a thermodynamic function or a variable as the case may be.

Similarly, the entropy S of a closed system is a useful function for characterizing the equilibrium or evolution of the system. But this quantity can also be considered as a thermodynamic variable.

In the set of state variables, there are two classes of variables:

– *Extensive variables*

An extensive variable is a variable depending on the size of the system (its extension):

m, amount of substance; N, number of moles; V, volume; U, internal energy; S, entropy.

– *Intensive variables*

An intensive variable is not a function of the size of the system, but rather of its position in space and time:

P, pressure; *T*, temperature; $\vec{E}$, electric field strength; $\vec{H}$, magnetic field strength.

1.3.1.4. *Conjugate variables*

Intensive and extensive variables are coupled two by two, restoring various forms of energy. Thus, the variables *P, V* are associated with the three-dimensional form of mechanical energy, the variables *T, S* with calorific energy.

Thermodynamic system	Intensive variable	Extensive variable
Mechanical **3D fluid system (gas)** **2D two-dimensional solid (film)** **1D two-dimensional solid (wire)**	P, pressure γ, surface tension F, force	V, volume A, area L, length
Electric (battery) continuous medium generator (solid)	V, voltage E, electric field strength	q, electric charge P, electric polarization
Magnetic	H, magnetic field strength	M, magnetic moment
Chemical	μ_m, chemical potential	m, mass of component

Table 1.1. *Conjugate state variables for simple thermodynamic systems*

Table 1.1 displays *conjugate state variables* for simple thermodynamic systems. This results in the state functions of a simple thermodynamic system. Among the state functions, one function is of particular interest. This is the *state equation*.

1.3.1.5. *State equation*

This equation, which results from experiments rather than from modeling, was originally obtained for gas in the general form:

$$f(P,V,T) = 0 \quad [1.17]$$

The work of Mariotte, Gay-Lussac and Charles led to the ideal gas equation:

$$PV = NRT \quad [1.18]$$

with R as the ideal gas constant = 8,314 J/K mole.

This state equation accounts for the real behavior of gases under low pressure.

Since then, many state equations have been studied and proposed in the literature, both for real gases and thermodynamic systems (see, for example, Feidt (2014, pp. 53–59)).

We will give an example of a type of state equation due to Van der Walls, in molar reference:

$$P = \frac{R \cdot T}{v - b} - \frac{a}{v^2} \quad [1.19]$$

with v as the molar volume.

a, b, R, are three parameters bound by the stresses at the critical point:

$$a = P_k v_k^2, \; b = \frac{v_k}{3}, \; R = \frac{8}{3} \cdot \frac{P_k v_k}{T_k} \quad [1.20]$$

where P_k, v_k, T_k are critical point coordinates.

In conclusion, for the state equations, even for a pure material, it is difficult (even impossible) to have a precise analytical expression, in particular that allows the expression of the variable chosen according to the others in all possible physical domains of the transformations. The most delicate zone remains close to the critical point, especially if the phase change is of interest, which will often be the case.

For a gas, the ideal gas equation remains limited.

The Van der Waals equation accounts for the phase changes of fluids.

1.3.2. *Thermodynamic transformation*

1.3.2.1. *Elementary thermodynamic transformation*

An elementary thermodynamic transformation accounts for the evolution of the system between two times t and $t + \Delta t$, with Δt time interval (duration) between two measurements. If Δt is small, the transformation can be studied with continuum mathematics: differential approach in time (dt tending to zero) by passing the limit.

For a finite transformation between an initial instant t_i and a final instant t_f, the evolution of a value $G(t)$ of the non-equilibrium system will be such that:

$$\Delta G(t_i, t_f) = \int_{t_i}^{t_f} dG(t) \qquad [1.21]$$

1.3.2.1.1. Quasi-static transformation

This transformation is described from a succession of thermodynamic equilibrium states accessible in the thermodynamic space often called the configuration space. The quasi-static transformation does not reveal the time variable or the associated kinematic, kinetic notions (velocity, acceleration, flow).

1.3.2.1.2. Quasi-staticity and relation to time

The evolution of equilibrium state i at the equilibrium state f occurs by successive stress relaxation. This assumes that after each elementary evolution, a new intermediate equilibrium state is reached, or by passing the limit it assumes sufficiently slow relaxation kinetics (velocity) to be able to skip intermediate equilibrium states (or to remain in their immediate vicinity). It depends on the *relaxation time* τ *of the system* with respect to the constraint considered.

This *time* τ must remain small compared to the stress relaxation time. If this condition is not met, the hypothesis of overall equilibrium (or even local) is no longer satisfied.

1.3.2.1.3. Example of expansion or compression of a gas in a cylinder

Only very slow movements of the piston in a cylinder can lead to quasi-static or even reversible processes. If the movement is rapid, the external forces or pressures vary rapidly, causing inhomogeneities in the cylinder. The entropy then increases during the transformation, which is irreversible, at most adiabatic.

This point will be taken up in this book in the chapter on finite speed thermodynamics (FST) (Petrescu et al 2015).

During expansion, there is a depletion of the gas in the vicinity of the piston. In compression, there is excess gas in this same area. Homogenization of the gas occurs by means of the flow until pressure equilibrium is attained in the cylinder. The pressure wave which travels at the speed of sound, a, in the gas interacts in a complex way with the walls and is dissipated in the form of heat.

The simplest estimate of an order of magnitude for the pressure wave relaxation time assumes full attenuation of the first interaction with the wall, whose characteristic length of the system is deemed to be $L = V^{1/3}$ (piston stroke). The relaxation time is written accordingly:

$$\tau = \frac{V^{\frac{1}{3}}}{a} \qquad [1.22]$$

If the expansion or compression time is larger than the characteristic time, then the quasi-static approximation makes sense.

1.3.2.1.4. The main thermodynamic transformations

Whether reversible or irreversible, the main thermodynamic transformations are:

– isochoric transformations ($V = c^{onst}$);

– isobaric transformations ($P = c^{onst}$);

– isothermal transformations ($T = c^{onst}$);

– adiabatic transformations (no heat exchange with the exterior).

However, in the latter case, it will be noted that the reversibility leads to an isentropic transformation ($S = c^{onst}$).

In addition, the transformations offer as a constraint either a constant intensive variable, or an extensive variable, or other constraints (Figure 1.4).

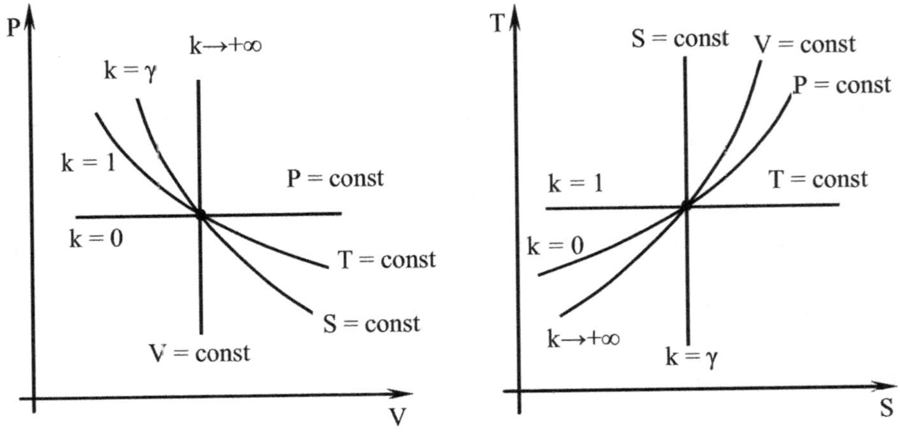

Figure 1.4. *Representation of thermodynamic transformations (Feidt 2014)*

1.3.2.1.5. Polytropic transformation

Figure 1.4 shows all the thermodynamic transformations in a Clapeyron diagram or an entropic diagram. For a three-dimensional thermomechanical system, this leads to the polytropic transformation equation:

$$PV^k = \text{const} \qquad [1.23]$$

with $k = \frac{N\gamma R + C(\gamma-1)}{NR+C(\gamma-1)}$;

$$\gamma = \frac{C_p}{C_v}$$

where C is the constant heat capacity with which the gas is in perfect thermal contact (infinite transfer conductance) and N is the number of moles of gas.

This relation accounts for the general thermodynamic behavior of the gas in a conservative transformation (mechanical and thermal equilibrium), but not adiabatic.

The polytropic exponent k is between the isothermal limit ($k = 1$) and the isentropic limit ($k = \gamma$).

This same transformation can be generalized (Feidt 2016, pp. 144–147) with the following for the polytropic exponent:

$$k = \frac{C_k - C_x}{C_k - C_y} \qquad [1.24]$$

with C_k as heat capacity of the transformation k of conjugate variables X (intensive) and Y (extensive), such that $\delta Q = C_k \, dT$.

1.3.2.2. *Transient and steady-state systems*

Section 1.3.2.1 presented quasi-static type thermodynamic transformations, and therefore non-functions of the time variable.

1.3.2.2.1. Transient transformation

The most general thermodynamic transformation is transient. It starts from an initial state at t_i to reach a final state at t_f, more or less irreversibly depending on the external constraints applied.

The associated modeling then reveals two fundamentally different types of derivative with respect to time: for a physical variable G, it is necessary to distinguish between:

– the capacitive term associated with $\frac{dG}{dt}$;

– the terms flows associated with $\dot{G}$.

Both have the same physical dimension but a different nature (see section 1.2.4).

The mathematical consequence is the obtaining of associated physical models in the form of differential systems to be solved, which is more complex than formulating the steady-state regime.

1.3.2.2.2. Stationary steady-state regimes

This regime, often referred to as a permanent regime by fluid mechanics, is a non-equilibrium regime: there are force gradients and correlatively associated flows. But the physical quantities do not depend on the moment at any point of the system.

The mathematical consequence is that the physical models are presented in the form of algebraic equations most often nonlinear, which is easier than solving differential systems.

1.3.3. *Energy transfers and conversion*

1.3.3.1. *Forms of energy and conversion*

Energy is a physical or multifaceted concept ubiquitous in nature and essential to life. The diversity of forms of energy induces many possibilities including:

– conversion from one form to another form;

– the transfer of a form of energy (particularly heat);

– the transport of a form of energy;

– storage in certain forms of energy.

To illustrate this notion, a brief matrix of energies is given in Table 1.2.

Form of energy	Mechanical	Electrical	Magnetic	Radiative
Mechanical		Mechano-electric		
Electrical	Electro-mechanical		Electromagnetic	Electrothermal
Magnetic	Magneto-mechanical	Magnetoelectric		
Radiative		Photovoltaic		
Chemical		Electrochemical		
Nuclear		Electronuclear		Nuclear physics
Calorific	Thermo-mechanical	Thermoelectric	Thermo-magnetic	Thermokinetics: radiation

Form of energy	Chemical	Nuclear	Calorific
Mechanical			Mechanothermal
Electrical	Electrochemical	Electronuclear	Electrothermal
Magnetic			Magnetothermal
Radiative	Photosynthesis		
Chemical			Thermochemistry
Nuclear			Thermonuclear
Calorific	Thermochemical		

Table 1.2. *Energy matrix*

The reader will be able to establish a list of commonly used converters as an exercise or even consider sub-matrices (case of mechanical–mechanical converters). Note in particular that the transfer phenomena remain limited to the diagonal of the matrix.

The books in this series will mainly focus on thermomechanical conversion and heat transfer.

1.3.3.2. *Generalized thermodynamic transformations*

By observing the forms of energy from a thermodynamic point of view, we can see that these are subject to:

– first law of thermodynamics or conservation law (including energy);

– second law of thermodynamics or evolution law, introducing the irreversibility of any real transformation.

This results in a generalized expression of the elementary work under the expression of a differentiable form:

$$\delta W = \sum_i Y_i \mathrm{d}X_i \qquad [1.25]$$

with Y_i as the generalized strength and X_i as the generalized shift, and this independently of the form of energy, except calorific energy.

Calorific energy also takes on the expression of a differential form:

$$\delta Q = T\mathrm{d}S \qquad [1.26]$$

On the other hand, note that dS is a total differential. The above expression of δQ corresponds to an equivalent equation, introduced in the case of polytropic transformation. In general, any sensible heat transfer over heat capacity is written as:

$$\delta Q = C\,\mathrm{d}T \qquad [1.27]$$

This extensive quantity is indexed by the associated thermodynamic transformation (C_P for an isobar; C_v for an isochore).

For this extensive quantity, the characteristic quantity of the constituent that we find in the books of thermo-physical data is preferred, in accordance with (Feidt 2016):

$$\delta Q = mc_m \mathrm{d}T = Nc_N \mathrm{d}T \quad [1.28]$$

Here, we arrive at mass or molar specific heats. Here, as an example of finite heat capacity, is the case of high, then low temperature. A good approximation at high temperature is $C = \text{const}$ if $T >> \theta$, with θ, the Debye temperature of the material (Ma 2020).

On the other hand, for low temperatures, if $T << \theta$, $C(T)$ becomes proportional to T^n (n and θ are functions of the material). This formalism is useful in very low temperature physics.

1.3.3.3. *The basics of heat transfers*

To describe the transfers, it is necessary to rely on the notion of flow introduced in section 1.3.2.2.1. Here, the concept of heat transfer is discussed:

$$\dot{Q} = \frac{\delta Q}{dt} \quad [1.29]$$

Based on thermokinetics, there are three modes of heat transfer, which will be useful in the development of true thermodynamics, the subject of this book and the corresponding series.

1.3.3.3.1. Heat transfer by conduction

Fourier's law expresses the heat flow conducted through a boundary element $\mathrm{d}A$ in the form:

$$\dot{Q} = -\lambda\, dA \frac{dT}{dn} \quad [1.30]$$

where $\frac{dT}{dn}$ is the normal derivative of the temperature on the normal directed toward the interior of the element $\mathrm{d}A$. If $\frac{dT}{dn} > 0$, the flow is lost by the system.

The value λ here is the isotropic homogeneous thermal conductivity on the boundary of the medium support necessary for the transfer.

1.3.3.3.2. Heat transfer by convection

It also requires material support, but fluid in motion. *Convection is referred to as natural*, when the motion of the fluid is due to variations in the physical properties of the fluid with temperature.

Convection is said to be forced, when the movement is caused by a mechanical device (fan; pump) forcing the circulation of the fluid.

The phenomenological law associated with convection is most often presented in the form of Newton's law of linear motion:

$$\dot{Q} = h \cdot dA \cdot \Delta T \qquad [1.31]$$

with *h* as the convective heat transfer coefficient per unit area. This coefficient is calculated from correlations available in the literature (Sacadura 1993). Note that ΔT is the temperature difference between the fluid and wall element.

1.3.3.3.3. Heat transfer by radiation

This mode of transfer does not require material support. It occurs between two separate surfaces by a medium transparent to radiation. Then the radiative transfer follows the Stefan–Boltzmann law, adjusted to take into account the mutual interactions:

$$\dot{Q} = \varepsilon_a \cdot dA \cdot (T^4 - T_A^4) \qquad [1.32]$$

with ε_a as hemispherical apparent emissivity, function of the geometry and nature of the facing surfaces; T_A as temperature of element d*A* (K); T as temperature of the surface exchanging with d*A* (K).

The books in this series discuss the important influence of heat transfer laws. Here, it is sufficient to say that the three modes of transfer can act in a coupled way.

Section 1.3.5 will discuss the important coupling between transfer and conversion in thermodynamics.

1.3.4. *Generalization of cycle and efficiency concepts*

1.3.4.1. *Cycles*

1.3.4.1.1. General definition of a simple cycle

A cycle consists of a succession of thermodynamic transformations either elementary, comprising a continuum, or well identified: see the Clapeyron diagram in Figure 1.5 for some examples of classic cycles.

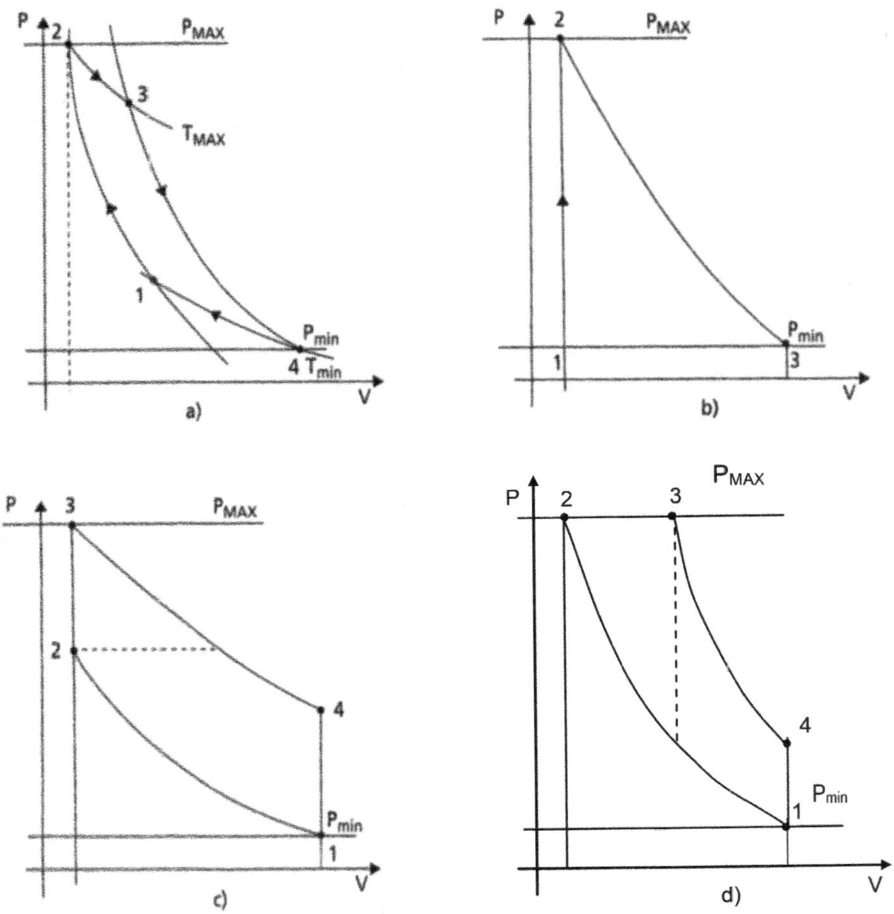

Figure 1.5. *Theoretical cycles: (a) Carnot, (b) Lenoir, (c) Beau de Rochas and (d) diesel (Feidt 2016)*

1.3.4.1.2. Some examples

Figure 1.5 shows that many cycles in practice include four successive transformations, but this is not necessarily always the case.

In addition to the representation in the form of the Clapeyron diagram, very clear from a mechanical point of view, it is necessary to compare the representation in the form of an entropy diagram (*T*, *S*) better adapted for thermodynamic reasoning.

Simple cycles will only be mentioned here. It is, however, also possible to consider nested cycles and cascading cycles. The potential is therefore very large. This raises the question of the choice of cycle: this is based on the concept of efficiency.

1.3.4.2. *Concept of efficiency*

The concept of efficiency is central in the choice from a physical, technical, economic and environmental point of view. This will be considered in detail in the fourth book in the series.

A general definition can be obtained from the system diagram in Figure 1.6.

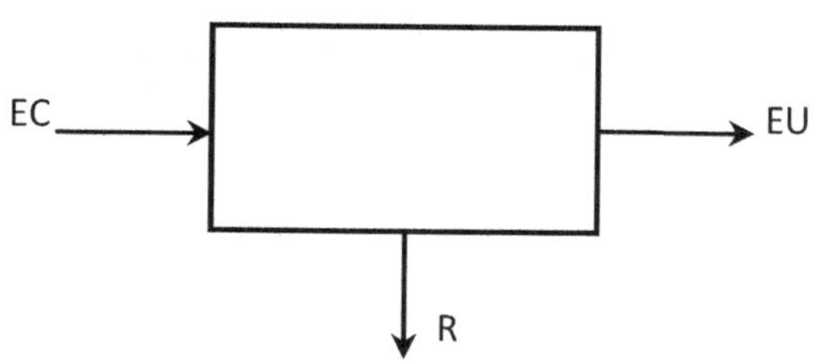

Figure 1.6. *Simple diagram of a system*

The useful effect *UE* is obtained after incurring an energy consumption *EC*. Correlatively, we see the appearance of heat rejections *R,* which could be recoverable or even detrimental.

In accordance with the first law of thermodynamics, we then define the efficiency:

$$\text{Efficiency} = \frac{UE}{EC} \quad [1.33]$$

This efficiency comes in a dimensionless form; this is important because it does not depend on the size of the system.

The distinctive characteristic is typical for energy. It is known as efficiency; in this case, energy efficiency, or more precisely efficiency in the sense of the first law η_I.

The example of a thermomechanical engine leads to:

$$\eta_I = \frac{W}{Q_{HS}} \quad [1.34]$$

with W as the mechanical energy produced, Q_{HS} as the amount of heat needed to produce W, as seen in section 1.2.3.

But an efficiency in the sense of the second law of thermodynamics can also be defined as follows:

$$\eta_{II} = \frac{W}{W_{rev}} \quad [1.35]$$

This efficiency reflects the importance of irreversibilities in the system studied.

An exergy efficiency η_{Ex} can also be defined, such that:

$$\eta_{Ex} = \frac{W}{\left(1-\frac{T_0}{T_{HS}}\right)Q_{HS}} \quad [1.36]$$

The reference for the exergy is then the exergy associated with the heat consumption.

These three examples clearly show the care that must be taken in studying the effectiveness of a system, depending on the desired result. This justifies the existence of the fourth book in this series.

This chapter will close with an illustration of the emblematic case of the Carnot engine.

1.4. Case study: the Carnot engine

1.4.1. *Energy and entropy balances*

The Carnot engine cycle is represented in its endoreversible version in Figure 1.7 in the temperature–entropy diagram.

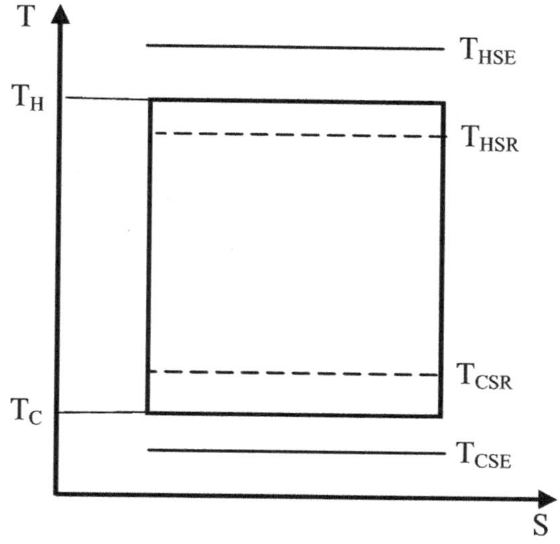

Figure 1.7. *Endoreversible representation of the Carnot engine*

The cycle which takes place between the temperatures T_H and T_C for the cycled fluid is traversed in the clockwise direction for the engine. This occurs in the opposite direction for a receiver (refrigerator or heat pump).

This diagram shows that the real engine imposes a heat source at a temperature higher than T_H. The hot source is represented here as a thermostat (infinite reservoir of heat at temperature T_{HSE}. The cold source corresponds to a thermostat at temperature T_{CSE}. The temperature differences of $\Delta T_H = (T_{HSE} - T_H)$ and $\Delta T_C = (T_{CSE} - T_C)$ are necessary for the heat transfers to take place.

Thus, the temperature staggering must satisfy the following inequality constraints for the engine:

$$T_{CSE} < T_C < T_H < T_{HSE}$$

The same reasoning applied to the receiver, with the same algebraic convention, gives:

$$T_C < T_{CSR} < T_{HSR} < T_H$$

With the additional assumption of an adiabatic system (without heat loss), for the fluid cycle we obtain:

– the energy balance:

$$W + Q_H + Q_C = 0 \qquad [1.37]$$

– the entropy balance (general case of an irreversible cycle: see Figure 1.8):

$$\frac{Q_H}{T_H} + \frac{Q_C}{T_C} + \Delta S_I = 0 \qquad [1.38]$$

or

$$\Delta S_H + \Delta S_C + \Delta S_I = 0 \qquad [1.39]$$

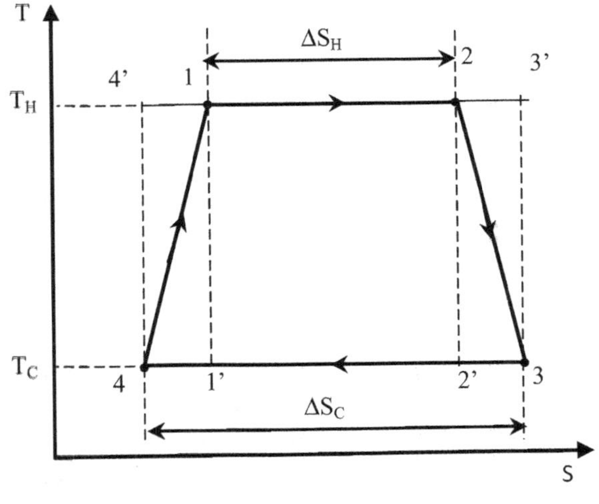

Figure 1.8. *Representation of an irreversible cycle of the Carnot engine*

The combination of the previous relations then easily leads to:

$$-W = (T_H - T_C)\Delta S_H - T_C \Delta S_I. \qquad [1.40]$$

This approach centered on the converter exposes the intensive variables (T_H, T_C), then the extensive variables (ΔS_H, ΔS_I).

It is important to notice that the two extensive variables have completely different roles and meanings. Therefore, ΔS_H is *transfer entropy*. This transfer entropy at the hot end of the converter appears as the natural reference of transfer entropies for an engine. It will be denoted ΔS. Then, ΔS_I corresponds to the *production of entropy* due to internal irreversibilities of the converter (mainly due to degradation of fluid and solid mechanical energy). If the engine is endoreversible, $\Delta S_I = 0$.

The system-centric approach (converter + source + sink) uses the assumption of adiabaticity in the form:

$$Q_H = T_{HS}\Delta S_{HS}; \quad Q_C = T_{CS}\Delta S_{CS} \qquad [1.41]$$

The energy balance of the system then becomes:

$$-W = (T_{HS} - T_{CS})\Delta S_{HS} - T_{CS}\Delta S_S \qquad [1.42]$$

with ΔS_S as entropy production of the system according to the entropy balance of the system:

$$\Delta S_{HS} + \Delta S_{CS} + \Delta S_S = 0 \qquad [1.43]$$

The expressions [1.40] and [1.42] impose the following inequality constraints on the engine:

$$\Delta S = \Delta S_H > \frac{T_C}{T_H - T_C}\,\Delta S_I \qquad [1.44]$$

$$\Delta S_{HS} > \frac{T_{CS}}{T_{HS} - T_{CS}}\,\Delta S_S \qquad [1.45]$$

1.4.2. *Entropy production and energy efficiency*

1.4.2.1. *System entropy production*

The production of entropy in the converter (endoirreversibility) is represented by the parameter of irreversibility ΔS_I. In reality, this parameter of the model is transformed into a complex function, which would deserve at least an experimental study, as well as a thorough entropy analysis.

On the other hand, the entropy produced by the heat transfer at the hot end and at the cold end can be explained as follows:

– entropy production at the hot end ΔS_{IH}:

$$\Delta S_{IH} = \Delta S_H - \Delta S_{HS} = \Delta S_H \frac{T_{HS}-T_H}{T_{HS}} \text{ or } \Delta S_{HS} \frac{T_{HS}-T_H}{T_H} \quad [1.46]$$

– entropy production at the cold end ΔS_{IC}:

$$\Delta S_{IC} = \Delta S_{CS} - \Delta S_C = -\Delta S_C \frac{T_C-T_{CS}}{T_{CS}} \text{ or } -\Delta S_{CS} \frac{T_C-T_{CS}}{T_C} \quad [1.47]$$

The entropy production of the system ΔS_S can be deduced as follows:

$$\Delta S_S = \Delta S_H \frac{T_{HS}-T_H}{T_{HS}} + \Delta S_I - \Delta S_C \frac{T_C-T_{CS}}{T_{CS}} \quad [1.48]$$

With this form depending on the transfer entropies entering and exiting the converter, other forms can be associated, for example, those depending on the transfer entropies at the hot and cold sources. By adopting this for the reference transfer entropy (input of the converter), we obtain:

$$\Delta S_S = \Delta S_H \left[\frac{T_C}{T_{CS}} - \frac{T_H}{T_{HS}}\right] + \frac{T_C}{T_{CS}} \Delta S_I \quad [1.49]$$

This expression confirms the evolution of entropy production of the system according to the variables T_H, T_C, ΔS_H and parameter ΔS_I.

1.4.2.2. *Energy efficiency (first law)*

This concept resulting from mechanics was generalized by Carnot in his report in the form of a thermomechanical efficiency. Here, this is specified further in the form of an efficiency in the sense of the first law. Generally, efficiency is defined as the ratio of a useful effect *UE*, obtained from consumption *D* (in the case under consideration, *EC* energy consumption), that is:

$$\eta_I = \frac{UC}{EC} \quad [1.50]$$

In the case of the Carnot engine, this expression becomes (see section 1.2.3):

$$\eta_I = \frac{W}{Q_{HS}} \quad [1.51]$$

If the engine is not adiabatic (loss Q_L), then it becomes:

$$Q_{HS} = Q_H + Q_L \qquad [1.52]$$

$$\eta_I = -\frac{W}{Q_H} \cdot \frac{Q_H}{Q_{HS}} \qquad [1.53]$$

The first term on the right-hand side represents the thermomechanical conversion efficiency η_C, the second term is a transfer efficiency η_T, such that:

$$\eta_I = \eta_C \cdot \eta_T \qquad [1.54]$$

These last two efficiencies are less than 1.

Therefore:

$$\eta_C + \eta_T = \eta_{Tot} \leq 2 \qquad [1.55]$$

With regard to the best efficiency distribution η_C, η_T subject to the constraint above, a simple derivation shows that the maximum efficiency in the sense of the first law corresponds physically to the equipartition of efficiencies:

$$\eta_C^* = \eta_T^* = \frac{\eta_{Tot}}{2} \qquad [1.56]$$

Consequently:

$$\text{Max}\, \eta_I = \frac{\eta_{Tot}^2}{4} \qquad [1.57]$$

The efficiency η_{Tot} is linked to the degree of perfection of the system, and therefore to its implementation cost. This point will be discussed later in the book (thermo-economy).

The application of the model in section 1.4.1 makes it possible to find the expression of η_I in the form (adiabatic case):

$$\eta_I = \frac{T_H - T_C}{T_H} - \frac{T_C}{T_H}\frac{\Delta S_I}{\Delta S_H} \qquad [1.58]$$

This expression shows, in the first term on the right-hand side, the efficiency of the endoreversible converter. The latter only takes into account the irreversibilities of transfer through T_H and T_C. The second term on the right-hand side relates to the internal irreversibilities of the converter. These appear in particular through the irreversible coefficient $\Delta S_I / \Delta S_H$ introduced by Novikov (1957).

A second approach in the literature (Ibrahim et al. 1991) involves introducing an irreversibility ratio I. In this case:

$$I_H = \frac{\Delta S_C}{\Delta S_H} \geq 1 \quad [1.59]$$

$$I_C = \frac{\Delta S_H}{\Delta S_C} \leq 1 \quad [1.60]$$

The same approach as above then leads to the new expression of efficiency, as a decreasing function of I_H:

$$\eta_I = 1 - I_H \frac{T_C}{T_H} \quad [1.61]$$

1.4.2.3. *Specifics of the exoreversible but endoirreversible engine*

It is possible to continue the previous development for a system and not the converter. This will not be done here for the sake of brevity (and left to the reader's discretion as an exercise). The methodology instead will be shown in the case where the thermal contacts at the source and at the sink are perfect (exoreversibility), but where the converter is irreversible (endoirreversibility). So, in this case:

$$|W| = |W_{\text{rev}}| - T_{CS}\Delta S_I \quad [1.62]$$

The term $W_I = T_{CS}\Delta S_I$ represents the mechanical dissipation. The work W_{rev} is indeed the upper limit of conversion into mechanical energy. With the temperature T_{CS} being at room temperature, W_I is also the energy of the engine. Furthermore:

$$\eta_I = 1 - I_H \frac{T_{CS}}{T_{HS}} = 1 - I_H(1 - \eta_{\text{Carnot}}) \quad [1.63]$$

The Carnot efficiency is the maximum efficiency that is obtained for a totally reversible system.

1.4.2.4. *Some cases of irreversibility as a function of intensive and extensive variables*

For the exoreversible engine, the expression of the mechanical energy in section 1.4.1 becomes:

$$-W = \Delta T_S \Delta S - T_{CS} \Delta S_I \qquad [1.64]$$

with $\Delta T_S = (T_{HS} - T_{CS})$ as the maximum intensive variable available for the engine.

The variable ΔS, on the other hand, is the extensive variable characteristic of the engine cycle. Proposed in this section is a short study of the influence of the production of entropy ΔS_I considered as a function of the characteristic variables of the engine, but also more generally, whatever the cycle (motor or receiver). Thereby, ΔS_I appears as a natural function of:

– ΔT_S: always positive intensive variable;

– ΔS: extensive reference variable, which can be positive (in the case of engines) or negative (in the case of heat pumps).

1.4.2.4.1. Simple case where the production of entropy depends only on extensity ΔS

With the second law of thermodynamics imposing $\Delta S_I > 0$, the simplest mathematical form associated with this case is the following quadratic form:

$$\Delta S_I = C_{IS} \Delta S^2 \quad \text{true } \forall\, \Delta S > 0 \text{ and } \Delta S < 0 \qquad [1.65]$$

Therefore, for the engine it becomes:

$$-W = \Delta S(\Delta T_S - C_{IS}\, T_{CS} \Delta S) \qquad [1.66]$$

The mechanical energy is then canceled out for $\Delta S = 0$ (absence of a cycle), but also for $\Delta S = \frac{\Delta T_S}{C_{IS}\, T_{CS}}$, this value tends toward infinity when C_{IS} tends toward zero (the reversible case allows all transfer entropies without limiting the engine).

The existence of an energy optimum then appears for a real irreversible engine. This optimum occurs for ΔS_E^* such that the derivative of W in relation to ΔS is canceled out, that is:

$$\Delta S_E^* = \frac{\Delta T_S}{2C_{IS}\,T_{CS}} \qquad [1.67]$$

There is a corresponding $\text{Max}_1 W_{exoreversible}$ whose expression is:

$$\text{Max}_1 W_{\text{exoreversible}} = \frac{\Delta T_S^2}{4C_{IS}\,T_{CS}} \qquad [1.68]$$

At the same time, Q_H^* equals:

$$Q_H^* = \frac{T_{HS}\,\Delta T_S}{2C_{IS}\,T_{CS}} \qquad [1.69]$$

The efficiency at maximum energy is:

$$\eta_I\big(\text{Max}_1(-W_{\text{exoreversible}})\big) = \frac{1}{2}\frac{\Delta T_S}{T_{HS}} \qquad [1.70]$$

that is half of the Carnot efficiency. This new result is remarkable.

1.4.2.4.2. Simple case where the production of entropy depends only on intensity ΔTS

For the same reasons as in the previous case, the validity of a positive entropy production ΔS_I leads to the simplest quadratic form (valid both for the engine case and for the receiver, given the algebraic convention adopted):

$$\Delta S_I = C_{IT}\Delta T_S^2 \qquad [1.71]$$

For the engine, this becomes:

$$-W = \Delta T_S(\Delta S - C_{IT}\,T_{CS}\Delta T_S) \qquad [1.72]$$

Again, the formal symmetry with the previous case leads to an optimum of mechanical energy for ΔT_S^* such that:

$$\Delta T_S^* = \frac{\Delta S}{2\,C_{IT}\,T_{CS}} \qquad [1.73]$$

The maximum energy corresponding to ΔS, the existence parameter of the constant cycle, equals:

$$\text{Max}_1(-W_{\text{exoreversible}}) = \frac{\Delta S^2}{4C_{IS}\,T_{CS}} \qquad [1.74]$$

The efficiency associated with this maximum energy again takes the same value as in the previous case (half of the Carnot efficiency).

1.4.2.4.3. Simple case where the production of entropy inversely depends on the duration of the cycle, Δt

In this case, ΔS_I takes the form:

$$\Delta S_I = \frac{C_{It}}{\Delta t} \qquad [1.75]$$

This form is perfectly compatible with the quasi-static limit restoring the reversibility of the converter, $\forall C_{It}$, which is a finite parameter. The energy of the engine then takes the form:

$$-W = \Delta T_S \Delta S - T_{CS} \frac{C_{It}}{\Delta t} \qquad [1.76]$$

This relation confirms that the optimization in this case relates to the mean power objective function $-\bar{\dot{W}}$, with the corresponding variable time period Δt (or frequency, $1/\Delta t$):

$$-\dot{W} = \frac{\Delta T_S \Delta S}{\Delta t} - \frac{C_{It} T_{CS}}{\Delta t^2} \qquad [1.77]$$

The engine operation imposes a minimum period in the interval $\Delta t \in \left[\frac{C_{It} T_{CS}}{\Delta T_S \Delta S}, \infty\right[$.

The average power optimum is obtained for Δt^* such that:

$$\Delta t^* = \frac{2C_{It} T_{CS}}{\Delta T_S \cdot \Delta S} = \frac{2C_{It} T_{CS}}{W_{rev}} \qquad [1.78]$$

with

$$\text{Max}_1(-W_{\text{exoreversible}}) = \frac{W_{rev}^2}{4C_{It} T_{CS}} \qquad [1.79]$$

The efficiency associated with this maximum average power again takes the same value as for the two previous cases, that is, half the Carnot efficiency:

$$\eta_I\left(\text{Max}_1(-W_{\text{exoreversible}})\right) = \frac{1}{2}\frac{\Delta T_S}{T_{HS}} = \frac{\eta_{\text{Carnot}}}{2} \qquad [1.80]$$

1.5. First conclusions and perspectives

This chapter has made it possible to resituate the evolution of thermodynamics since the work of Carnot and his applications, in particular in the field of thermomechanical engines.

The compelling concepts discussed made it possible to revisit and complete previous studies in an original way:

– fundamental role of transfer entropy, which is essential in the coupling of heat sources and sinks with the converter;

– existence of a trade-off between transfer phenomena and energy conversion;

– concept of (energy) efficiency which Carnot intuitively generalized starting from mechanics; this generalization and this concept will be discussed in more detail in another book in this series;

– extension of the textbook case, Carnot engine, to any type of engine. Here again, a book on this topic will discuss the approaches discussed in this chapter, but in a more complete way and closer in technique and technology;

– extension of the case study to the receiver: a book will discuss the concepts and applications specific to the field of refrigerators and heat pumps.

This introductory chapter has addressed the central problem of controlling the production of entropy resulting from the unavoidable irreversibilities of our world and universe.

The simple overall approach proposed has made it possible to show the existence of the energy optimum of the Carnot engine, which appears due to the existence of the irreversibilities:

– maximum energy depending on variables: extensive (ΔS, transfer entropy), intensive (ΔT_S, engine gradient or thermodynamic force);

– average maximum power as a function of a time variable (Δt, cycle period or inverse of the frequency). Note that time and its related arrow are inseparable from the second law of thermodynamics, known as the law of evolution. However, it is specified here that the reported example rather refers to a stationary steady-state regime than to operation in an unsteady or intermediate regime. This point will be discussed later;

– efficiency according to the first law at the previous maxima.

Then, in a systematic way, for the three cases studied, the same efficiency is obtained in the sense of the first law at the optimum, which equals half of the Carnot efficiency whatever the optimized extent (energy or power: flux of extensity):

$$\eta_I(\text{optimum}) = \frac{\eta_{\text{Carnot}}}{2}$$

A conjecture of universality of this result can thus be suggested.

The extensions to these conclusions are therefore all mapped out and will be developed throughout the books in this series. Aspects relating to the environment and the economy, essential to the engineer (see the last chapter of this book), will also be discussed.

1.6. References

Carnot, S. (1824). *Réflexion sur la puissance motrice du feu.* Bachelier, Paris.

Feidt, M. (1987). *Thermodynamique et optimisation énergétique des systèmes et procédés*, 1st edition. Lavoisier TEC & DOC, Paris.

Feidt, M. (2013). *Thermodynamique optimale en dimensions physiques finies.* Hermes, Paris.

Feidt, M. (2014). *Génie energétique. Du dimensionnement des composants au pilotage des systèmes*. Dunod, Paris.

Feidt, M. (2016). *Energétique. Concept et applications*. Dunod, Paris.

Ibrahim, O.M., Klein, S.A., Mitchell, J.W. (1991). Optimum heat power cycles for specified boundary conditions. *J. Eng. Gas Turb. Power*, 113(4), 514–521.

Idel'çik, I.E. (1986). *Mémento des pertes de charges*. Eyrolles, Paris.

Ma, Y.H. (2020). Effect of finite size heat source's heat capacity on the efficiency of heat engine. *Entropy*, 22(9), 1002.

Novikov, I.I. (1957). Efficiency of an atomic power generation installation. *The Soviet Journal of Atomic Energy*, 3(11), 1269–1272.

Petrescu, S., Costea, M., Feidt, M., Ganea, I., Boriaru, N. (2015). *Advanced Thermodynamics of Irreversible Processes with Finite Speed and Finite Dimensions. A Historical and Epistemological Approach, with Extension to Biological and Social Systems*. AGIR, Bucharest.

Sacadura, J.F. (1993). *Initiation aux transferts thermiques*, 2nd edition. Lavoisier TEC & DOC, Paris.

2

Two-Heat-Source Thermodynamic Cycles: Representation as a Ternary Diagram

Julien RAMOUSSE
LOCIE, UMR 5271, CNRS, Université Savoie Mont Blanc, Bourget du Lac, France

2.1. Introduction to two-heat-source systems

Energy conversion systems have experienced a real boom since the discovery of the steam engine in the 18th century. These external combustion engines ensure the transformation of the thermal energy of water vapor (produced by a boiler) into mechanical energy and they largely contributed to the first industrial revolution.

These discoveries were accompanied by the emergence of a new science, dedicated to the study of energy transformation processes: *thermodynamics*. The etymology comes from the Greek words: *thermos* for heat and *dynamis* for movement. Thus, thermodynamics initially focused on any transformation involving thermal and mechanical energy (Joule 1850; Clausius 1851; Carnot 1897). It was subsequently generalized to all systems mobilizing energy (in any form: mechanical, electrical, chemical, thermal energy), including fluids, technical systems, living organisms and large atmospheric systems (meteorological) to planetary systems. This science was then able to address non-equilibrium systems (thermodynamics of

Advances in Thermodynamics and Circular Thermoeconomics,
coordinated by Michel FEIDT and Antonio VALERO-CAPILLA.

irreversible processes) (Haase 1968; Prigogine 1968; Glansdorff 1971; Zahra 1989) and opened up to other disciplines such as economics or evolution.

On our scale, many technical applications of these principles have emerged, the main ones being:

– engines (internal or external combustion) used for propulsion, or electrical generation, involving different thermodynamic cycles. These systems allow the conversion of thermal energy (generally resulting from combustion) into work (mechanical or electrical energy). The main applications encountered in our daily lives are internal combustion engines (gasoline and diesel engines), steam turbines and generators;

– *heat pumps* allowing thermal energy to be transferred from an environment at low temperature to an environment at high temperature, thus reversing the natural direction of the spontaneous transfer of thermal energy. Depending on the objective, these devices can ensure the heating of a hot environment (heat pump), or the cooling of a cold environment (refrigerator). When the purpose of the device is both to heat a hot environment and cool a cold environment simultaneously, the system is then qualified as a double-function heat pump. These installations are also very common in our daily lives: refrigerators, air conditioners and various heating systems.

All of the aforementioned systems are qualified as two-heat-source systems, because they involve (at least) two heat reservoirs at different temperature levels. It is thus possible to distinguish the heat reservoirs by their temperature level: a hot reservoir at a high temperature T_h and a cold reservoir at a low temperature T_c, with $T_c \leq T_h$. The temperature difference between the two reservoirs determines the maximum performance of conversion of work into heat (or vice versa), as detailed below.

2.2. Definitions and convention

2.2.1. *Work and heat*

Energy exists in different forms as mechanical energy, electrical energy, chemical energy and thermal energy. Because of their difference in quality, we will distinguish *work*, denoted as W, from thermal energy (or *heat*), denoted as Q indexed h or c, depending on the temperature of the heat reservoir (hot or cold) involved for this heat exchange. Chemical energies will not be discussed here because these energies are not involved in the two-heat-source systems considered here. Energy exchanges for a two-heat-source system are therefore limited to the work exchanged W, and heat exchanged Q_h and Q_c with hot and cold heat reservoirs, respectively.

In the case of a steady-state system, the powers exchanged by the system will be preferentially used. The work power will then be noted $\dot{W}$ and the thermal power $\dot{Q}$. And, in the case of a fluid in a closed cycle, the specific (mass) energies used are denoted as w and q for the specific work and the specific heat exchanged, respectively.

In general, a two-heat-source system can thus be described using the diagram proposed in Figure 2.1, where the system and the energy exchanges are represented (work exchanged with the environment and heat exchanged with the heat reservoirs).

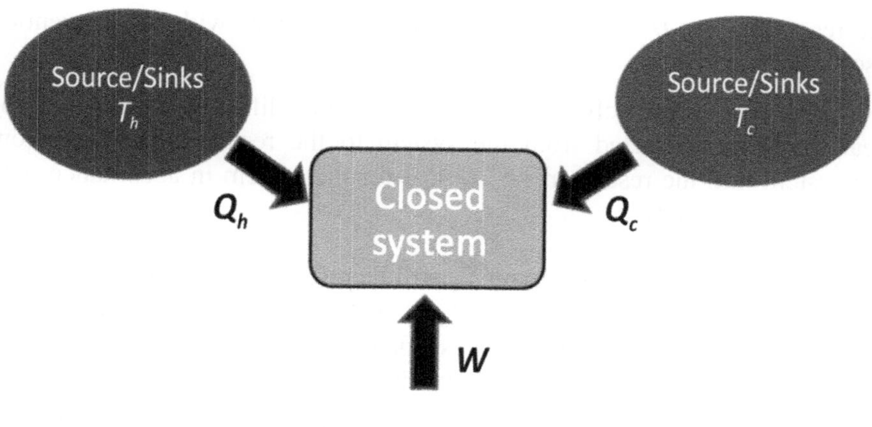

Figure 2.1. *Two-heat-source system. For a color version of this figure, see www.iste.co.uk/feidt/advances.zip*

2.2.2. Sign convention

By convention, the energies exchanged by the system are counted algebraically:

– positively, for an energy received by the system;

– negatively, for an energy released by the system.

This convention applies to both work energy and heat energy exchanged by the system. The sign of the energy flow therefore indicates the direction in which this flow is exchanged.

It should be mentioned that other conventions are sometimes applied (Moran 2014; Bejan 2016), which do not call into question the analysis proposed below. The signs of the energy flows exchanged just have to be adapted, depending on the convention chosen.

2.2.3. *Heat source/sink*

For pedagogical reasons, heat reservoirs are distinguished according to the direction of the heat flow exchanged with them:

– *heat source*: the reservoir will be qualified as a source when the heat transfer is directed from the reservoir to the system (heat withdrawal from the reservoir, i.e., $Q > 0$ for the system in accordance with the convention used);

– *heat sink*: conversely, the reservoir will be qualified as a sink when the heat transfer is directed from the system to the reservoir (of heat from the system into the reservoir, i.e., $Q < 0$ for the system in accordance with the convention used).

The high-temperature reservoir T_h will be referred to as a "hot" reservoir, and the low-temperature reservoir T_c as a "cold" reservoir, with $T_c \leq T_h$. The reservoirs are assumed to be sufficiently large in size so that heat exchanges with the system do not modify their temperature. As a result, reservoir temperatures are assumed to be constant, with $\theta = {}^{T_c}/_{T_h} \in [0; 1]$.

In the same way, the system will be qualified as a receptor when it receives work ($W > 0$) and a generator when it produces work ($W < 0$).

2.2.4. *First-law balance*

The energy balance, resulting from the first law of thermodynamics, applied to a unit of mass describing a cycle of a two-heat-source system reduces to the following equality:

$$q_h + q_c + w = 0 \quad [2.1]$$

This balance is written from the specific energies (mass) in order to generalize its application to the case of a two-heat-source system in a closed cycle and to the case of a two-heat-source system in an open cycle in steady-state. In the first case, the energies involved will be deduced from the specific energies by multiplication of the constant mass of working fluid m involved: $Q_h + Q_c + W = 0$. In the second case, the first law can be written using the thermal and work powers involved by multiplying the specific energies by the constant flow rate of the working fluid $\dot{m}$: $\dot{Q}_h + \dot{Q}_c + \dot{W} = 0$.

This equation corresponds to the equation of a plane in the orthonormal three-dimensional reference frame $(\overrightarrow{q_h}, \overrightarrow{q_c}, \vec{w})$. Consequently, all the two-heat-source systems can be represented on plane *P*, which will be introduced later, for a simplified graphical interpretation.

2.2.5. *Second-law balance*

2.2.5.1. *General case*

The second law of thermodynamics introduces limitations to the functioning of these two-heat-source systems, expressed by the Clausius inequality:

$$\frac{q_h}{T_h} + \frac{q_c}{T_c} \leq 0 \qquad [2.2]$$

The equality is only satisfied in the ideal case of a reversible engine (the so-called Carnot system, detailed in the next section). More generally, this inequality can be translated using the following equality, by introducing specific entropy generation $\sigma \geq 0$, defined positive:

$$\frac{h}{T_h} + \frac{q_c}{T_c} + \sigma = 0 \qquad [2.3]$$

with $\sigma \geq 0$.

Thus, the irreversibilities of the system expressed by the positive definite entropy generation (equal to zero in the case of an ideal reversible system) reduce the possible operating conditions of two-heat-source systems by specifying the impossible transformations (that would imply $\sigma < 0$).

The consequences of this second law on the operating conditions of two-heat-source systems in the proposed graphical representation will be detailed next.

2.2.5.2. *Reversible case (Carnot system)*

The study of the reversible operation of two-heat-source systems is interesting because it allows for qualifying the ideal operating conditions of these systems. This ideal case can only be reached with the following reversible transformations, describing the Carnot cycle represented in Figure 2.2:

– adiabatic and reversible compression (isentropic);

– isothermal heating at T_h (heat exchange with hot reservoir);

– adiabatic and reversible expansion (isentropic);

– isothermal cooling at T_c (heat exchange with cold reservoir).

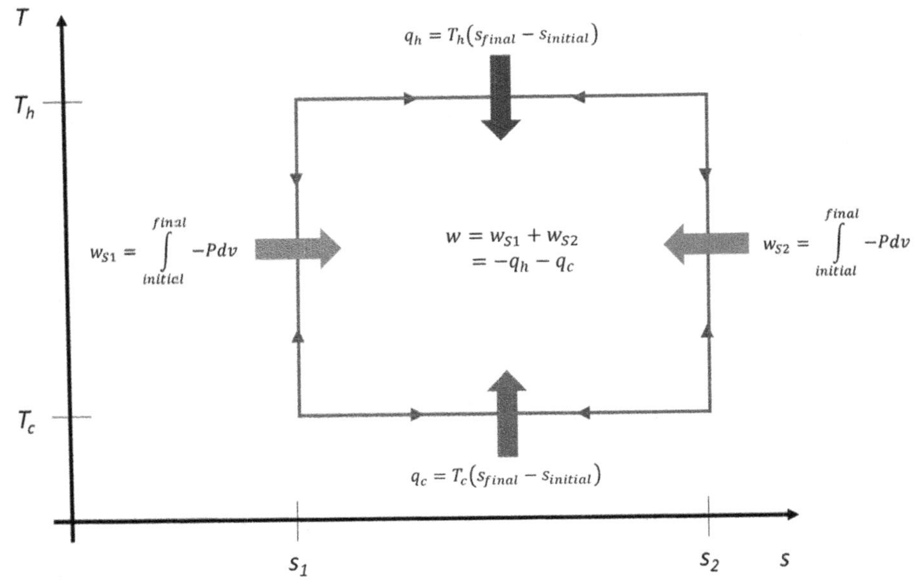

Figure 2.2. *Temperature–entropy diagram of the Carnot cycle. For a color version of this figure, see www.iste.co.uk/feidt/advances.zip*

Note that this cycle can function in both directions:

– clockwise: reversible engine cycle. In this case, $q_h > 0$, $q_c < 0$ and $w < 0$;

– anti-clockwise: reversible heat pump cycle. In this case, $q_h < 0$, $q_c > 0$ and $w > 0$;

Depending on the direction of the cycle, the heating of the working fluid ($q > 0$) occurs at T_h (engine cycle) or at T_c (heat pump cycle). Conversely, the cooling of the working fluid ($q < 0$) occurs at T_c for the engine cycle or T_h for the heat pump cycle. Similarly, the isentropic compression ($w > 0$) and relaxation ($w < 0$) phases are reversed according to the direction of cycle.

Under these reversible operating conditions ($\sigma = 0$), the Clausius inequality (equation [2.2]) then becomes:

$$\frac{q_h}{T_h} + \frac{q_c}{T_c} = 0 \qquad [2.4]$$

Thus, the ratio of the source temperatures directly imposes the ratio of the specific heats exchanged by the system with its reservoirs:

$$\frac{q_f}{q_c} = -\frac{T_f}{T_c} = -\theta \qquad [2.5]$$

2.2.5.3. *Performance*

The energy performance of these conversion systems is generally characterized by their efficiency (in engine mode) or their coefficient of performance (COP) in heat pump mode, defined as follows:

$$\text{Engine system } \eta = \frac{-w}{q_h} \qquad [2.6]$$

$$\text{Heat pump } COP_h = \frac{q_h}{w} \qquad [2.7]$$

$$\text{Refrigerator } COP_c = \frac{q_c}{w} \qquad [2.8]$$

The performance coefficients of heat pump systems are dissociated according to the desired useful effect (heat pump: high-temperature heating from a cold source, or refrigerator: low-temperature cooling with heat release at high temperature). The term double-function heat pump is used when the two effects (cooling at low temperature and heating at high temperature) occur simultaneously.

One can easily verify, using the first law (equation [2.1]), that $COP_h = COP_c + 1$.

A consequence of the first and second laws (equations [2.1] and [2.4]) is that the maximum specific work exchanged by a reversible two-heat-source system can be linked to the specific heat exchanged with the hot or cold reservoir, which then makes it possible to define the Carnot efficiency for the engine mode and the Carnot COP for a heat pump:

$$\eta^H = \frac{-w}{q_h} = 1 - \frac{T_c}{T_h} \quad [2.9]$$

$$COP_h^C = \frac{q_h}{w} = \frac{T_h}{\Delta T} \quad [2.10]$$

$$COP_c^C = \frac{q_c}{w} = \frac{T_c}{\Delta T} \quad [2.11]$$

These definitions clearly highlight the dependence of the performance of two-heat-source systems on the temperature levels of the heat reservoirs with which they exchange. In order to compare the performance of two two-heat-source systems that exchange heat with reservoirs at different temperatures, it may be relevant to compare their efficiency according to the second law, defined as follows:

For an engine system: $\varepsilon = \frac{\eta}{\eta^C}$ [2.12]

For a heat pump: $\varepsilon = \frac{COP}{COP^C}$ [2.13]

2.3. Graphic representations – state of the art

The equilibrium states of a pure compound can be plotted on a diagram 3D (*P*-*v*-*T*), as proposed in the work of Gibbs (Gibbs 1873). This representation made it possible to establish 2D phase diagrams, classically used for the study of the thermodynamic transformations according to two state variables of the system (*P*-*v* diagram (Clapeyron 1960), h-*s* chart (Mollier 1904) and *T*-*s* (temperature-entropy) diagram) and commonly used to represent thermodynamic cycles (Lhuillier 1992; Pérez 2001; Moran

2008; Borel 2010; Bejan 2016), as introduced in the previous section for the Carnot cycle. However, these representations only allow an indirect evaluation of the energy flows involved in the cycle. It is thus useful to propose additional graphical tools, which allow the study of two-heat-source thermodynamic systems, allowing a heat-work conversion (and vice versa), with a view to better understand the operation of these systems and their performances. Among these approaches, graphical representations are of particular interest because they guarantee a visual interpretation accessible to a large audience. Thus, several graphical representations have been proposed in the literature to help facilitate the analysis of two-heat-source systems. This section therefore proposes a state of the art of graphical tools dedicated to the analysis of these systems in order to later introduce the representation in the ternary diagram.

2.3.1. *Q-T diagram*

A representation of the operation of two-heat-source systems in engine and heat pump modes in an orthogonal *Q-T* diagram (Heat-Temperature) was proposed in Bejan (2016), as presented in Figure 2.3.

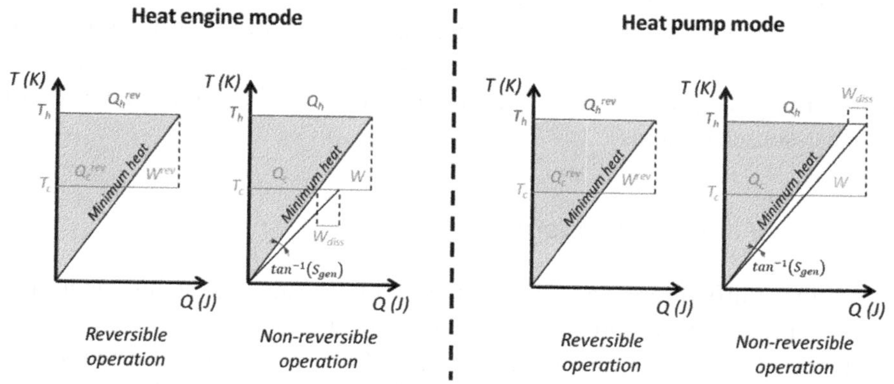

Figure 2.3. *Q-T diagram. Adapted from Bejan (2016). For a color version of this figure, see www.iste.co.uk/feidt/advances.zip*

Note here that the signs of the energy flows exchanged by the system are not considered (absolute values), which requires duplicating this representation in the case of engine and heat pump modes. The reversible operation of two-heat-source systems imposes a linear dependence of the specific heat fluxes as a function of the temperature of the reservoirs (equation [2.5]), represented by the "minimum heat" line. The reversible work exchanged by the system is deduced from the difference in the heat flows exchanged, depending on the temperature of the reservoirs (positive in engine mode, and negative in heat pump mode). In the case of non-reversible operation, an additional amount of heat must be dissipated by the system due to irreversibilities (W_{diss}), which results in a deviation from reversible operation. This dissipated work (positive according to the convention adopted in this chapter) is transmitted to the heat sink exchanging with the system (cold reservoir in the engine mode or hot reservoir in the heat engine mode) and results in a reduction in the work available in heat pump mode, and an increase in the work required in heat pump mode. Note here that this additional energy dissipated is expressed directly as a function of the entropy generation of the system.

This representation has the advantage of illustrating the deviation from the reversible operation of a real two-heat-source system. The additional energy dissipated in the heat sink due to irreversibilities can thus be evaluated. However, by relying on the absolute values of the energy flows exchanged, this approach requires the implementation of two distinct representations depending on the operating mode (and therefore of the sign of the energy flows exchanged) studied. Later it is shown that by taking into account the sign of the energy flows exchanged, the proposed representation in the ternary diagram allows a unified analysis of all the operating modes of two-heat-source systems.

2.3.2. *Borel and Favrat diagram*

Another unified representation of the two-heat-source systems' operation has been proposed by Borel and Favrat (2010). In order to qualify the exergy efficiency of these systems, it is assumed that one of the two reservoirs with which the system exchanged is at ambient temperature T_a (reference temperature for exergy analysis). The second reservoir considered may therefore have a temperature T_b higher or lower than T_a. It is thus possible to represent the energy efficiency (equations [2.6]–[2.8]) and the exergy efficiency of any two-heat-source system in an orthogonal basis whose axes are defined by the ratio of the temperatures of the reservoirs ${}^{T_b}/_{T_a} \in [0; +\infty]$ and the ratio of the work exchanged to the absolute value of the heat exchanged with the reservoir at temperature T_b, ${}^{w}/_{|q_b|} \in [-\infty; +\infty]$, as presented in Figure 2.4.

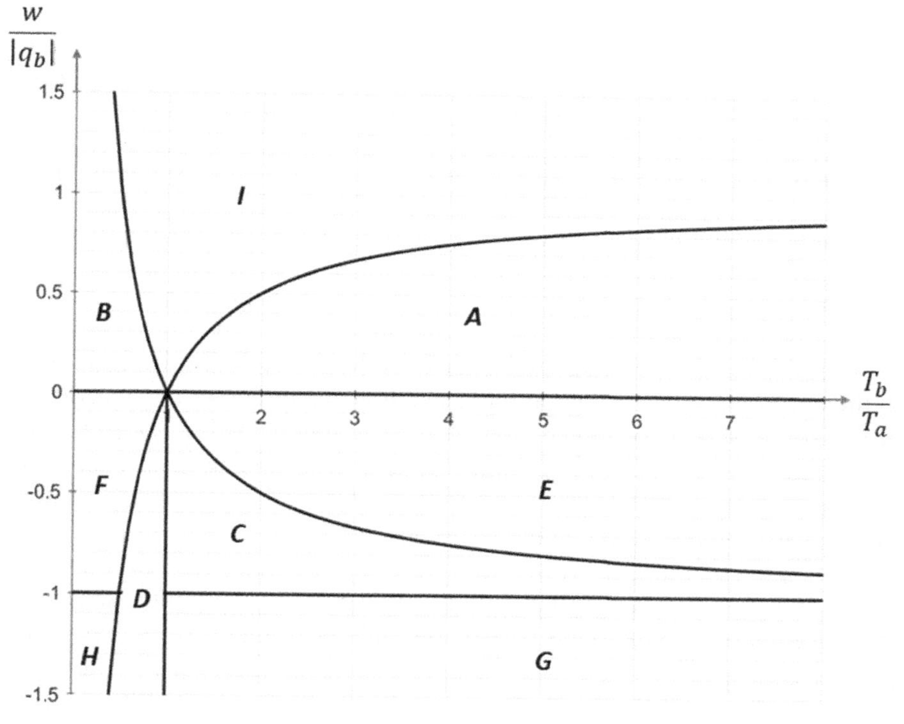

Figure 2.4. *Graphical representation by Borel and Favrat (2010) of the energy efficiency of two-heat-source cycles*

This representation of the operating conditions of two-heat-source systems makes it possible to identify and qualify different operating modes, represented by the areas noted from A to I in the diagram in Figure 2.4:

– Domain A: engine cycles (with $T_0 = T_c$) => *Engine Mode*

– Domain B: engine cycles (with $T_0 = T_h$) => *Engine Mode*

– Domain C: heat pump cycles => *Heat Pump Mode*

– Domain D: refrigeration cycles => *Heat Pump Mode*

– Domain E: dissipative cycles => *Forced Heat Transfer Mode*

– Domain F: dissipative cycles => *Forced Heat Transfer Mode*

– Domain G: dissipative heating cycles => *Heat Dissipation Mode*

– Domain H: dissipative cycles => *Heat Dissipation Mode*

– Domain I: domain impossible to reach => *Impossible*

These different operating modes, detailed below, will also be highlighted in the proposed ternary diagram representation, according to the equivalence proposed. It should be noted here that heat pump cycles ($T_b \geq T_a$) and refrigeration cycles ($T_b \leq T_a$), both characterizing operation in heat pump mode according to the desired useful effect, can be distinguished according to the relative temperatures of the reservoirs.

The reversible operating conditions correspond to the points located on the two equilateral hyperbolas passing through the point (1,0) and having the vertical abscissa ${}^{T_b}/_{T_a} = 0$ and the horizontal coordinates ${}^{w}/_{|q_b|} = \pm 1$ as asymptotes.

This representation has the advantage of offering a unified analysis of the operating conditions of two-heat-source systems. However, the dissipative modes, although identified, are unfortunately not discussed and described in more detail. Also, assuming that one of the reservoirs is at room temperature T_a (which can therefore play the role of the low or high temperature reservoir) results in a subdivision of the operation in engine mode distributed over two distinct and non-adjacent domains (Domains A and B). However, this condition is necessary to introduce the exergy efficiencies of these systems (not established in the proposed representation).

The ternary diagram representation proposed in the following section makes it possible to overcome these discontinuities, which can impair the graphical interpretation of these systems by proposing a unified and continuous representation of the operating conditions of two-heat-source systems, close to the representation of Raveau, presented below.

2.3.3. *Raveau diagram*

Raveau (2021), Pérez (2001) and Lhuillier (1995) propose to represent the functioning of two-heat-source systems in an orthonormal basis with the axes $(Q_h; Q_c)$, as presented in Figure 2.5. A generalization of this approach in the case of multiple heat reservoirs has been proposed by Alefeld and Radermarcher (1993).

The application of the first law of thermodynamics (equation [2.1]) makes it possible to deduce the axis W, defined by the bisector of the axes Q_h and Q_c. Note here that even if this axis W can be identified, it is not standardized with respect to the axes Q_h and Q_c. Although this representation is very close to the representation proposed here, the ternary diagram representation overcomes this lack with the introduction of three standardized non-orthogonal axes Q_h, Q_c and W, which allows a deeper graphical interpretation via the introduction of polar coordinates.

Furthermore, it is possible to distinguish the different operating modes of two-heat-source systems, according to the signs of the energies exchanged by the system. The different operating modes, which are detailed below, are also highlighted in the ternary diagram representation proposed here. It should be noted here that this representation has the advantage of a unified and continuous graphical analysis of all the operating conditions of two-heat-source systems, in contrast to previous representations.

Finally, the Carnot limit (imposed by the second law for the reversible case, equation [2.4]) can be plotted with a straight line passing through the origin, whose slope is set by the temperature of the reservoirs with which the system exchanges. This limit implies a limitation of the operating conditions achievable by two-heat-source systems (reduced to the half-plane to the left of the Carnot limit), which will also be identified in the ternary diagram.

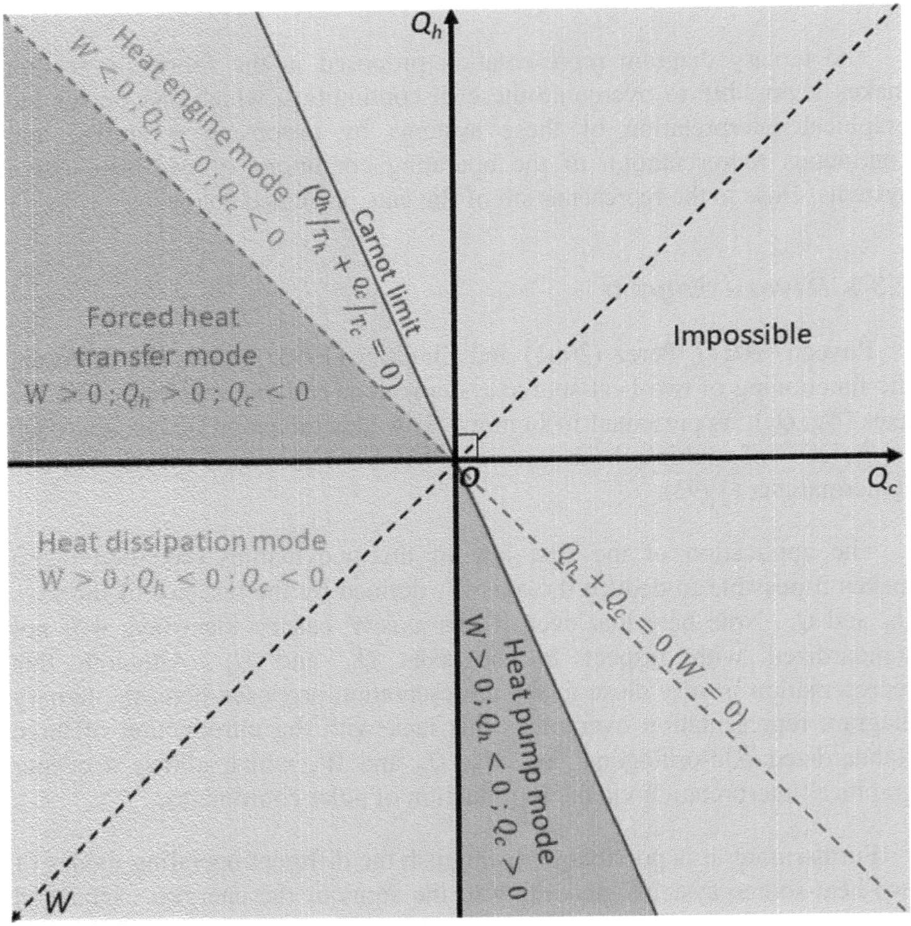

Figure 2.5. *Raveau diagram. For a color version of this figure, see www.iste.co.uk/feidt/advances.zip*

2.4. Ternary diagram

The ternary diagram representation (Ramousse 2016, 2018) presented here therefore aims to complete and improve the graphical representations previously mentioned in order to facilitate the graphical analysis of two-heat-source system performance.

2.4.1. *Ternary diagram representation*

2.4.1.1. *Introduction to the ternary diagram*

The first-law balance, expressed by equation [2.1], suggests the equation of a plane P in a three-dimensional orthonormal basis $(q_h'; q_c'; w')$ that it therefore seems relevant to characterize. This plane, orthogonal to the trisector of the orthonormal axes $(q_h'; q_c'; w')$, can be represented in Figure 2.6 where the axes $(q_h,; q_c; w)$ are deduced by projection of the axes $(q_h'; q_c'; w')$ in the plane. The axes $(q_h,; q_c; w)$ are standardized and are related by a rotational symmetry of $^{2\pi}/_3$ with the origin $M°(0,0,0)$ for the center of rotation.

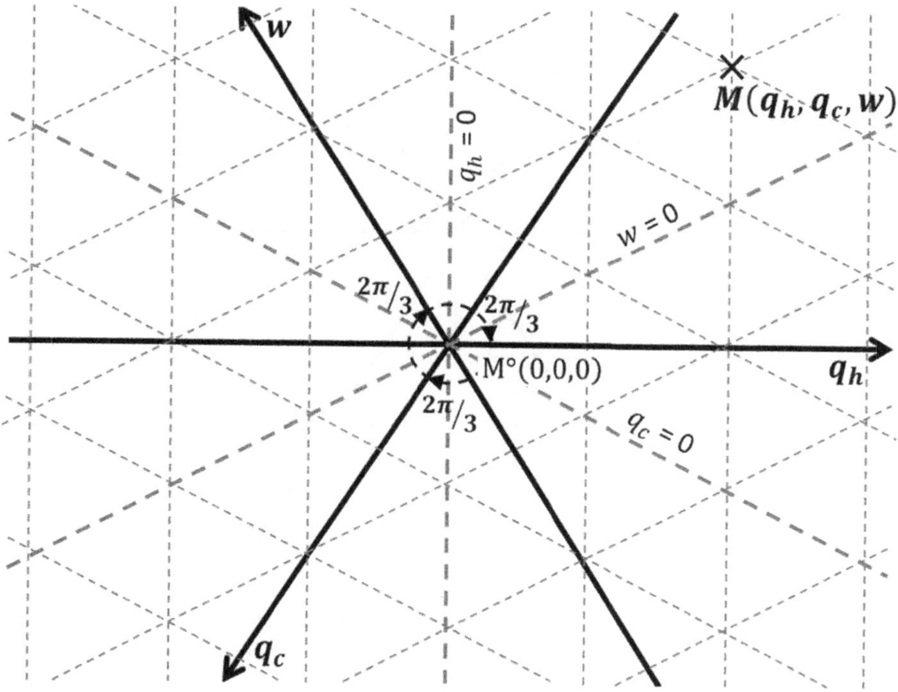

Figure 2.6. *Construction of the ternary diagram*

At any point $M(q_h, q_c, w)$ in the plane, the first-law balance (equation [2.1]) is verified. Thus, any two-heat-source system can be represented in the proposed ternary diagram from the knowledge of only two of the three specific energies (q_h, q_c and w) exchanged between the system and its environment; the unknown energy can be directly read off the diagram.

2.4.1.2. *Modes of operation*

As presented in Figure 2.7, six regions (or sextants) can be identified on the P plane, depending on the signs of the energy flows exchanged (q_h, q_c and w). Each of these regions corresponds to a distinct mode of operation, as detailed below. Regions are arbitrarily numbered for ease of interpretation.

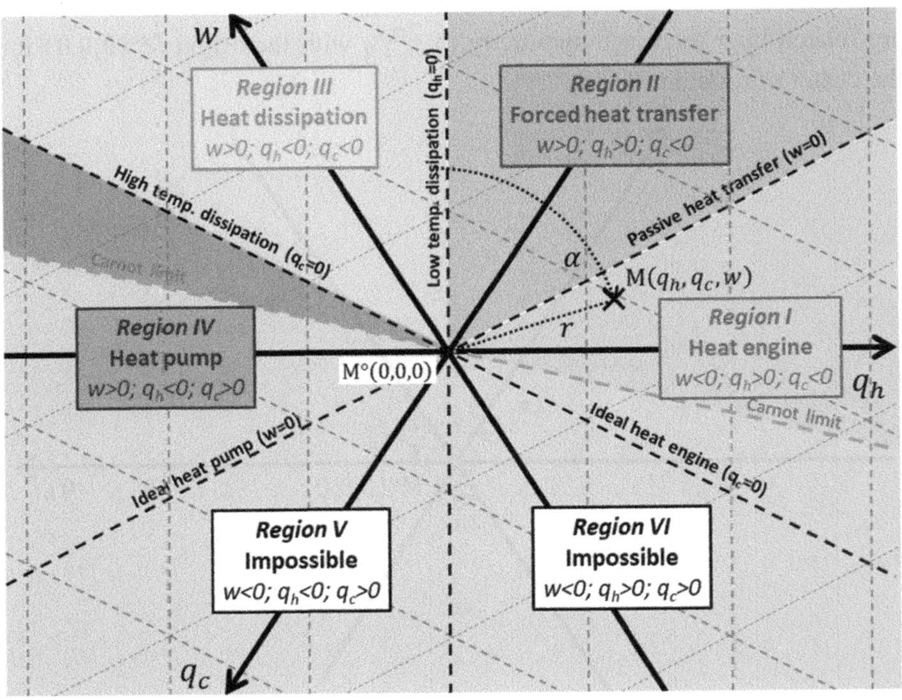

Figure 2.7. *Modes of operation. For a color version of this figure, see www.iste.co.uk/feidt/advances.zip*

– *Region I: Heat engine* $\mathbf{w < 0, q_h > 0, q_c < 0}$

This is the only mode where the system operates as a generator ($w < 0$) to produce work from a heat transfer from a high-temperature heat source to a low-temperature heat sink. From an exergy point of view, the system converts part of the high-quality thermal exergy taken from a hot source into work by rejecting low-quality thermal exergy to a cold sink.

– *Region II: Forced heat transfer* $\mathbf{w} > 0$, $\boldsymbol{q_h} > 0$, $\boldsymbol{q_c} < 0$

In this mode of operation, the system uses the work received to intensify the heat transfer from a high-temperature heat source to a low-temperature heat sink. This mode of operation is conventionally used in the case of heat exchangers, where the work supplied to the system is used to counterbalance the pressure drops of the heat transfer fluids. In this case, the exergy associated with work and high-temperature heat is dissipated as low-temperature thermal exergy at the cold sink.

– *Region III: Heat dissipation* $\mathbf{w} > 0, \boldsymbol{q_h} < 0, \boldsymbol{q_c} < 0$

In this case, the system converts the work received into heat, which it transfers to the hot and cold sinks. No heat source is involved in this operating mode. The work exergy received by the system is thus dissipated as thermal exergy at the two heat sinks of different temperatures.

– *Region IV: Heat pump* $\mathbf{w} > 0, \boldsymbol{q_h} < 0, \boldsymbol{q_c} > 0$

In heat pump mode, the system uses the work received to transfer heat from a cold heat source to a hot heat sink. A distinction is commonly made between the heating heat pump and cooling heat pump, depending on the temperatures of the heat reservoirs and the desired useful effect (heating or cooling), even if these two effects are simultaneous. When the two effects are simultaneously valorized, these systems are then qualified as double-function heat pumps. In this mode, the work exergy received by the system makes it possible to raise the exergy level from a low-temperature heat source to a high-temperature heat sink.

– *Region V: Impossible* $\mathbf{w} < 0, \boldsymbol{q_h} < 0, \boldsymbol{q_c} > 0$

In accordance with the second law of thermodynamics (equation [2.2]), this mode of operation is not possible because the temperatures of the reservoirs are distinct and positive ($T_h \geq T_c \geq 0$). It is not possible to spontaneously transfer heat from a low-temperature heat source to a high-temperature heat sink without external work input.

– *Region VI: Impossible* $\mathbf{w} < 0, \boldsymbol{q_h} > 0, \boldsymbol{q_c} > 0$

As before, this mode of operation is not possible due to the second law of thermodynamics (equation [2.2]), because the temperatures of the reservoirs are positive ($T_h \geq T_c \geq 0$). A low-temperature heat sink is indeed necessary for any conversion of thermal energy into work.

Note here that in line with the first law of thermodynamics (equation [2.1]), no region corresponds to the following operating conditions (not appearing in the plane P):

– $w > 0, q_h > 0, q_c > 0$

– $w < 0, q_h < 0, q_c < 0$

The origin of the plane $M°(q_h = q_c = w = 0)$ is characterized by no exchange of energy between the system and its environment: the system is at thermodynamic equilibrium.

The realistic modes of operation of two-heat-source systems are therefore reduced to regions I to IV. In addition to the two potentially reversible operating modes mainly studied (heat engine mode – region I; heat pump mode – region IV), two additional dissipative modes are identified (forced heat transfer mode – region II; heat dissipation mode – region III). Depending on its position in the ternary diagram, the mode of operation of any two-heat-source system $M(q_h, q_c, w)$ can therefore be directly deduced.

2.4.1.3. *Boundaries*

The proposed representation has the advantage of continuity between the different operating modes previously introduced. The interfaces between two adjacent operating modes in the ternary diagram are characterized by the cancellation of one of the energy flows exchanged and can therefore be qualified as follows:

– *Boundary between regions VI/I*: at this interface $q_c = 0$, therefore implying $q_h = -w$, meaning a total conversion of the thermal energy taken from the high-temperature heat source into work. By passage to the limit, this operating condition could only be reached in the case of an ideal reversible operating engine system with a heat sink at absolute zero temperature.

– *Boundary between regions I/II*: the operating conditions at this boundary are characterized by $w = 0$, that is, $q_h = -q_c$. These conditions correspond to passive heat transfer by conduction, free convection or radiation.

– *Boundary between regions II/III*: at this interface $q_h = 0$, therefore implying $w = -q_c$. In this case, all the work received by the system is transferred to the low-temperature heat sink, corresponding to low-temperature heat dissipation.

– *Boundary between regions III/IV*: here, $q_c = 0$, therefore implying $w = -q_h$. In this case, the work received by the system is transferred to the high-temperature heat sink, corresponding to high-temperature heat dissipation.

– *Boundary between regions IV/V*: as for the boundary between regions I and II, these operating conditions correspond to passive heat transfer ($w = 0$ and $q_h = -q_c$). By passage to the limit, these operating conditions can only be reached by an ideal operating heat pump where the work provided is zero (in the case of reservoirs at equal temperatures).

The characterization of these interfaces between the different adjacent regions validates the continuity of the proposed representation.

2.4.1.4. *Carnot limit*

Note here the temperature of the reservoirs has not yet been considered. However, as introduced previously, reversible operation defines a limit to the operation of two-heat-source systems, which is expressed as a function of the temperature of the reservoirs with which the system exchanges heat (equation [2.4]). This limit implies a proportional relationship between the heat flows exchanged between the system and its reservoirs (equation [2.5]). This equality results in a straight line passing through the origin in the proposed ternary diagram, as presented in Figure 2.7, whose slope is imposed by the ratio of the temperatures of the reservoirs θ. Since the reservoir temperatures are positive with $T_c \leq T_h$, then $0 \leq \theta \leq 1$, the Carnot limit only appears on the potentially reversible modes (heat engine and heat pump modes, corresponding to regions I and IV, respectively) for which $-1 \leq {}^{q_c}/_{q_h} \leq 0$. Indeed, regions II and III correspond to dissipative operating modes: they cannot be reached by an equivalent reversible system (which would then imply $\theta > 1$ in region II because ${}^{q_c}/_{q_h} \leq -1$, or $\theta < 0$ in region III because ${}^{q_c}/_{q_h} \geq 0$).

This Carnot limit thus reduces the possible operating conditions of two-heat-source systems as a function of the reservoir's temperature, as described by the Clausius inequality (equation [2.2]). Taking into account irreversibilities, via the positive specific entropy generation σ, then implies that the operating conditions of a two-heat-source system for a given reservoir temperature are limited to the half-plane above the Carnot limit in the proposed representation. The influence of reservoir temperatures on the achievable operating conditions of two-heat-source systems are thus linked to the slope of the Carnot limit, which defines the allowable half-plane for the system considered, depending on these reservoir temperatures.

2.4.2. *Interpretation using polar coordinates*

Although the analysis of the operating conditions can be based on the standardized axes (q_h, q_c and w), the introduction of polar coordinates (r, α) into the proposed representation opens the way to a complementary and simplified graphical interpretation of the performance of two-heat-source systems.

As all the operating conditions of two-heat-source systems can be represented in the previously introduced plane P, any operating condition can be described using the polar coordinates defined by the radius r, the radial distance from the considered point at the origin $M°(0,0,0)$, and the angle α between the perpendicular to the axis q_h and the point under consideration, as represented in Figure 2.7.

2.4.2.1. *Energy intensity*

Trigonometric analysis enables expressing the radius r as a function of the energy flows exchanged between the system and its environment as follows:

$$r^2 = {q_h}^2 + \left(\frac{q_c - w}{\sqrt{3}}\right)^2 = {q_c}^2 + \left(\frac{w - q_h}{\sqrt{3}}\right)^2 = w^2 + \left(\frac{q_h - q_c}{\sqrt{3}}\right)^2 \qquad [2.14]$$

More generally, the radius r can be expressed as the quadratic sum of the energy flows entering and leaving the system (in accordance with the sign convention adopted: positive, for incoming flows and negative for outgoing flows):

$$r^2 = {e_{ent}}^2 + {e_{sor}}^2 \qquad [2.15]$$

with $e_{ent/sor}$ corresponding either to the only energy flux exchanged in the direction considered, or to the difference in the fluxes exchanged divided by $\sqrt{3}$ when two energy flows are exchanged in the same direction. The coefficient $\sqrt{3}$ comes from the angle $^{2\pi}/_{3}$ between axes q_h, q_c and w.

So, the radius r can be interpreted as the energy intensity involved at the operating point considered. The larger the radius, the greater the energy flows exchanged.

By applying the second law of thermodynamics (equation [2.3]), the radius can also be expressed as a function of the reservoir's temperature, the specific heat exchanged with the hot (or cold) reservoir and the specific entropy generation as follows:

$$\begin{aligned} r^2 &= \frac{4}{3} q_h^{\,2\left[1-\theta+\theta^2-\frac{T_c\sigma}{q_h}\left(1-2\theta-\frac{T_c\sigma}{q_h}\right)\right]} \\ &= \frac{4}{3} q_c^{\,2}\left[1-\frac{1}{\theta}+\frac{1}{\theta^2}-\frac{T_h\sigma}{q_c}\left(1-\frac{2}{\theta}-\frac{T_h\sigma}{q_c}\right)\right] \end{aligned} \quad [2.16]$$

In the case of reversible operation (Carnot system, $\sigma = 0$), this expression reduces to:

$$r^{c^2} = \frac{4}{3} q_h^{\,2}(1-\theta+\theta^2) = \frac{4}{3} q_c^{\,2}\left(1-\frac{1}{\theta}+\frac{1}{\theta^2}\right) \quad [2.17]$$

2.4.2.2. Conversion efficiency

Trigonometric analysis enables linking angle α to the energy flows exchanged between the system and its environment using the following expressions:

$$\sin\alpha = \frac{q_h}{r} \text{ and } \cos\alpha = \frac{w-q_c}{\sqrt{3}r} \quad [2.18]$$

As shown in Figure 2.7, the previously introduced regions can be described using the angle α. Thus, its value makes it possible to directly identify the operating mode of the system considered:

- $\alpha \in \left[^{\pi}/_{3}; ^{2\pi}/_{3}\right]$ Region I – Engine Mode
- $\alpha \in \left[0; ^{\pi}/_{3}\right]$ Region II – Forced Heat Transfer Mode

$-\alpha \in [-\pi/3; 0]$ Region III – Heat Dissipation Mode

$-\alpha \in [-2\pi/3; -\pi/3]$ Region IV – Heat Pump Mode

Moreover, the symmetry suggested above leads to the proposal of the following definitions for the evaluation of the conversion performance of these systems, depending on the nature (source or sink) of the high-temperature heat reservoir.

– **High-temperature heat source ($q_h > 0$)**

In the case of a high-temperature heat source supplying heat to the system ($q_h > 0$), the classic definition of the conversion efficiency of engine systems is retained and can be expressed as follows:

$$\eta_{q_h+} = \frac{-w}{q_h} \qquad [2.19]$$

Based on this definition, it is possible to verify that any two-heat-source system operating in heat engine mode (region I – $q_h > 0,\ q_c < 0,\ w < 0$) has a positive efficiency lower than 1. This efficiency tends toward 1 in the case of a heat engine system in ideal operation ($q_c = 0$, i.e., $q_h = -w$, boundary between regions VI/I) and decreases continuously to the value of 0 in the case of passive heat transfer ($w = 0$, i.e., $q_h = -q_c$, boundary between regions I/II).

And any system operating in forced heat transfer mode (region II – $q_h > 0,\ q_c < 0,\ w > 0$) presents a negative efficiency, as expected for a dissipative mode. The efficiency decreases from 0 in the case of a system operating in passive heat transfer mode (boundary between regions II/III) down to $-\infty$ for a system operating in low-temperature heat dissipation mode (boundary between regions II/III).

By application of the second law of thermodynamics (equation [2.3]), this conversion efficiency can be expressed as follows:

$$\eta_{q_h+} = 1 - \theta - \frac{\sigma T_c}{q_h} \qquad [2.20]$$

In the case of reversible operation ($\sigma = 0$), the Carnot efficiency of a two-heat-source system exchanging with a high-temperature heat source is then defined by:

$$\eta_{q_h+}{}^C = 1 - \theta = 1 - \frac{T_c}{T_h} \qquad [2.21]$$

It is then easy to verify that for any real system ($\sigma \geq 0$), exchanging with a high temperature source ($q_h > 0$) has a conversion efficiency lower than the Carnot efficiency: $\eta_{q_h+} \leq \eta_{q_h+}{}^C$.

– High-temperature heat sink ($\boldsymbol{q_c} < 0$)

The symmetry described above encourages us to introduce an original definition of the conversion efficiency of two-heat-source systems exchanging with a high-temperature heat sink based on the following observation: The energies exchanged by a system Σ^{HP} operating in heat pump mode ($Q_h{}^{HP} < 0$, $Q_c{}^{HP} > 0$, $W^{HP} > 0$) can be counterbalanced by an antagonistic system operating in heat engine mode, such as $Q_h{}^{Eng} = -Q_h{}^{HP} > 0$ $Q_c{}^{Eng} = -Q_c{}^{HP} < 0$ and $W^{Eng} = -W^{HP} < 0$. The operation of these two complementary systems therefore does not involve any exchange of work with their environment or heat energy with the heat reservoirs. It therefore seems natural to define the conversion efficiency of a system operating in heat pump mode as the complement to that of the conversion efficiency of a heat engine system. Thus, it is proposed to define the conversion efficiency of a two-heat-source system exchanging with a high-temperature heat sink as follows:

$$\eta_{q_h-} = 1 - \eta_{q_h+} = \frac{q_c}{-q_h} \qquad [2.22]$$

This efficiency can be interpreted as the ability of the system to transfer heat energy from a low-temperature heat source to a high-temperature heat sink. With this definition, the efficiency of any system operating in heat pump mode (region IV: $q_h < 0$, $q_c > 0$ and $w > 0$) is positive and less than 1. It is equal to 1 in the case of a heat pump in ideal operation ($w = 0$, i.e., $q_c = -q_h$, boundary between regions VI/V) and decreases to 0 in the case of a system in high-temperature heat dissipation mode ($q_c = 0$, i.e., $w = -q_h$, boundary between regions III/VI).

And, any system operating in heat dissipation mode (region III: $q_h < 0$, $q_c < 0$, $w > 0$) has negative efficiency, decreasing from zero for high-temperature heat dissipation mode ($q_c = 0$, i.e., $w = -q_h$, boundary between regions III/VI) down to –∞ for a system operating in low temperature heat dissipation mode (border between the regions II/III).

This original definition of the conversion efficiency of two-heat-source systems exchanging heat with a high-temperature heat sink can be easily linked to the definitions conventionally used to characterize the performance of heat pumps (COP_h for a heat pump or COP_c for a refrigerator) using the following expressions:

$$COP_h = \frac{-q_h}{w} = \frac{1}{1-\eta_{q_h-}} \qquad [2.23]$$

$$COP_c = \frac{q_c}{w} = \frac{\eta_{q_h-}}{1-\eta_{q_h-}} \qquad [2.24]$$

$$\eta_{q_h-} = \frac{COP_c}{COP_h} = \frac{COP_h-1}{COP_h} = \frac{COP_c}{COP_c+1} \qquad [2.25]$$

By application of the second law of thermodynamics (equation [2.3]), this efficiency can also be expressed as:

$$\eta_{q_h-} = \theta + \frac{\sigma T_c}{q_h} \qquad [2.26]$$

and in the case of reversible operation ($\sigma = 0$), the Carnot efficiency of a two-heat-source system exchanging with a high-temperature heat sink is then:

$$\eta_{q_h-}{}^C = \frac{COP^C{}_c}{COP^C{}_h} = \frac{T_c}{T_h} = \theta \qquad [2.27]$$

So, any real system ($\sigma \geq 0$) exchanging with a high-temperature sink ($q_h < 0$) has a conversion efficiency lower than the Carnot efficiency: $\eta_{q_h-} \leq \eta_{q_h-}{}^C$.

Applying these definitions in the proposed ternary diagram leads to representations of the conversion efficiency of two-heat-source systems presented in Figure 2.8. Note here the symmetry of the conversion efficiencies along the axis perpendicular to the axis q_h.

Through some analytical developments, it is possible to express the efficiency of the system as a function of the polar angle α previously introduced as follows:

$$\tan|\alpha| = \frac{\sqrt{3}}{1-2\eta_i} \qquad [2.28]$$

with $\eta_i = \eta_{q_c-}$ for $\alpha \in \left[-{}^{2\pi}/_{3}\,;0\right]$ and $\eta_i = \eta_{q_c+}$ for $\alpha \in \left[0\,;{}^{2\pi}/_{3}\right]$.

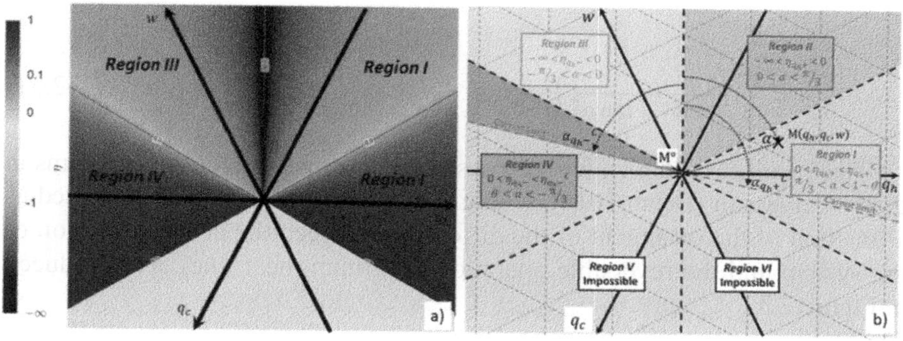

Figure 2.8. *(a) Conversion efficiency of two-heat-source systems. (b) Schematic representation of the conversion efficiency of two-heat-source systems as a function of α. For a color version of this figure, see www.iste.co.uk/feidt/advances.zip*

This expression confirms the axial symmetry noted above and illustrates the direct dependence between the conversion efficiency of the system (η_{q_c+} or η_{q_c-}) and the angle α relative to the operating point considered in the proposed diagram.

Accordingly, the polar angle can be expressed as a function of the specific entropy generation by the system and the reservoir temperatures:

$$\tan\alpha = \frac{\sqrt{3}}{2\theta-1+2\frac{\sigma T_c}{q_h}} \qquad [2.29]$$

In the case of a reversible system (Carnot system, $\sigma = 0$), this expression reduces to the following expression, defining the maximum value of the angle α^C depending on the reservoir temperatures:

$$\tan\alpha^C = \frac{\sqrt{3}}{2\theta-1} \qquad [2.30]$$

with $\alpha_{q_h+}{}^C \in [\pi/3 ; 2\pi/3]$ and $\alpha_{q_h-}{}^C \in [-2\pi/3 ; -\pi/3]$.

This result confirms the representation of the Carnot limit as a straight line passing through the origin, whose slope is imposed by the reservoir temperatures, with $\alpha_{q_h+}{}^C = \alpha_{q_h-}{}^C + \pi$. Obviously, any real system ($\sigma \geq 0$) verifies the condition: $|\alpha| \leq |\alpha^C|$.

The specific entropy generation can then be deduced from angles α and α^C by:

$$\frac{\sigma T_c}{q_h} = \frac{\sqrt{3}}{2}\left(\frac{1}{\tan\alpha} - \frac{1}{\tan\alpha^C}\right) \quad [2.31]$$

This result is in line with the representation of two-heat-source systems in the *Q-T* diagram (Bejan 2016), where the entropy generation is expressed as a function of the cotangent of the difference between the actual operation of the system and the reversible operation (minimum heat line), as introduced in Figure 2.3.

Thanks to the standardization of axes q_h, q_c and w (related by a rotational symmetry of $2\pi/3$ with the origin $M°(0,0,0)$ for the center of rotation), the graphical interpretation of the operating conditions of two-heat-source systems is simplified. Indeed, the introduction of the polar coordinates in the proposed ternary diagram thus allows an intuitive graphical analysis: the distance r from the considered operating point $M(q_h, q_c, w)$ to the origin $M°(0,0,0)$ is relative to the intensity of the energy flows exchanged between the system and its environment, and the angle α between the perpendicular to the axis q_c and the considered operating point $M(q_h, q_c, w)$ can be directly linked to its conversion efficiency, which have a value between 0 and 1 for potentially reversible operating modes (heat engine and heat pump modes) and negative for dissipative modes (forced heat transfer and heat dissipation modes) according to the definitions introduced previously.

2.5. Application examples

In order to illustrate the implementation and the interest of the proposed representation, this section endeavors to graphically characterize the operation of real irreversible two-heat-source systems using the proposed representation.

The ideal operation of two-heat-source systems according to the Carnot cycle assumes that the transformations of the system (isothermal and isentropic) are carried out in infinite time and space. Indeed, a transformation can only be considered reversible if it is quasi-static, then assuming a thermodynamic equilibrium at all times between the system and its environment, thus implying an infinitely slow transformation. Similarly, isothermal transformations assume infinitely small heat transfer, involving an infinitely large heat exchanger so that the powers exchanged do not result in a temperature rise.

In order to describe the actual operation of these conversion systems, it is possible to break down the causes of the irreversibilities in the system considered as follows:

– external to the system (exo-irreversible system), taking into account the irreversibilities linked to energy exchanges between the system and its environment;

– internal to the system (endo-irreversible system), taking into account the irreversibilities linked to the real transformations of the working fluid.

In a first approach (Ramousse 2018), it is thus possible to separately qualify the functioning of endo-reversible (exo-irreversible) systems and exo-reversible (endo-irreversible) systems in order to assess the influence of these different sources of irreversibility. The study of a real system will obviously have to include internal (endo-irreversible) and external (exo-irreversible) irreversibilities.

2.5.1. *Endo-reversible (exo-irreversible) two-heat-source systems*

First, let us consider an endo-reversible system (Carnot system) exchanging with heat reservoirs through specific thermal resistances ${R^{th}}_h$ and ${R^{th}}_c$ with the high-temperature and low-temperature reservoirs, respectively, as shown in Figure 2.9. These thermal resistances are related to the irreversibilities linked to the heat exchanges between the system and its associated reservoirs. The system thus operates at intermediate temperatures $T_h{}^i$ and $T_c{}^i$ related to the heat reservoir temperatures T_h and T_c by the following equations:

$$q_h = q_h{}^C = \frac{(T_h - T_h{}^i)}{R^{th}{}_h} \qquad [2.32]$$

$$q_c = q_c{}^C = \frac{(T_c - T_c{}^i)}{R^{th}{}_c} \qquad [2.33]$$

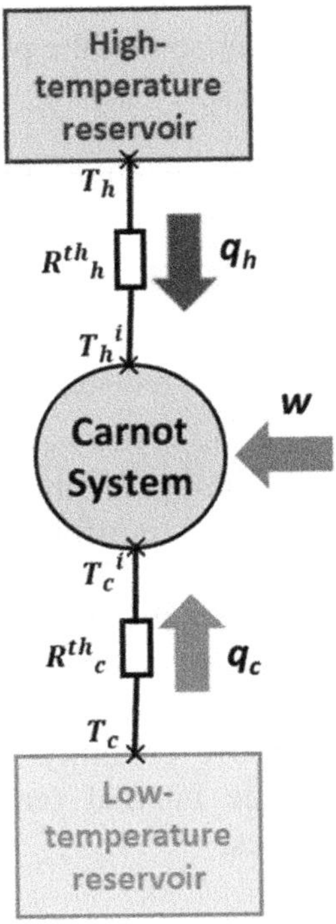

Figure 2.9. *Endo-reversible two-heat-source system. For a color version of this figure, see www.iste.co.uk/feidt/advances.zip*

In order to illustrate the influence of external irreversibilities linked to thermal contact resistances on the operation of two-heat-source systems, Figure 2.10 shows the operating points of an endo-reversible system (Carnot system) connected to the heat reservoirs at $T_h = 450$ K and $T_c = 300$ K, via equal specific thermal resistances $R^{th}{}_h = R^{th}{}_c$, varying from 1×10^{-3} to 5×10^{-3} $\text{K} \cdot \text{kg} \cdot \text{J}^{-1}$.

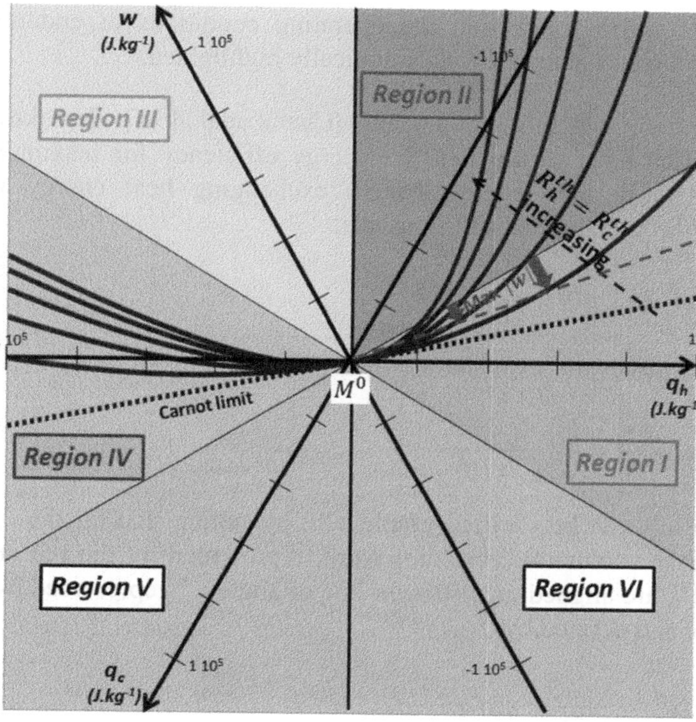

Figure 2.10. *Endo-reversible two-heat-source system for $T_h = 450\ K$ and $T_c = 300\ K$ ($\theta = \frac{2}{3}$), and $R^{th}{}_h = R^{th}{}_c$ varying from 1 × 10^{-3} to 5 × 10^{-3} $K.kg.J^{-1}$. The red dotted line represents the operating points at maximum work (in absolute value) produced by the system. For a color version of this figure, see www.iste.co.uk/feidt/advances.zip*

The influence of the thermal resistances between the system and its heat reservoirs is clearly highlighted in this figure. The lower the thermal resistance, the simpler the heat exchanges between the system and its reservoirs, thus making it possible to achieve high conversion efficiencies for higher energies exchanged. Conversely, in the case of high thermal resistances between the system and its reservoirs, the conversion efficiencies of the system are very quickly reduced when the energy exchanges increase. Consequently, when the specific heat received by the system from the heat source becomes too great, the system can no longer operate in heat engine mode and can only operate in forced heat transfer mode. The influence of external irreversibilities, resulting from thermal resistances between the system and its reservoirs, on the operating conditions of endoreversible two-heat-source systems are thus graphically highlighted.

In addition, different approaches (Chambadal 1957; Novikov 1958; Curzon 1975) have shown that the energy efficiency for maximum work production of a heat engine system exchanging heat energy with its reservoirs through thermal resistances is:

$$\eta_{q_h+}{}^{N} = 1 - \sqrt{\theta} \qquad [2.34]$$

The maximum work exchanged under these conditions is:

$$\mathrm{Max}|w_{<0}| = T_h \frac{\left(1-\sqrt{\theta}\right)^2}{R^{th}{}_h + R^{th}{}_c} \qquad [2.35]$$

This result can be verified graphically by noting that all the operating conditions at maximum (absolute) work, represented by the red arrows in Figure 2.10, are aligned on a straight line of angle α^N represented by the red dotted line and defined by:

$$\tan \alpha^N = \frac{\sqrt{3}}{1-2\eta_{q_h+}{}^{N}} = \frac{\sqrt{3}}{2\sqrt{\theta}-1} \qquad [2.36]$$

In the example shown in Figure 2.10, for $\theta = {}^2/_3$, the efficiency at maximum work is $\eta_{q_h+}{}^{N} = 0.1835$ and the corresponding angle α^N is 1.2204 rad.

2.5.2. *Exo-reversible (endo-irreversible) two-heat-source systems*

Now let us consider Carnot, Stirling and Ericsson exo-reversible cycles (without thermal resistances between the system and reservoirs) of an ideal gas. Only these cycles are considered here for illustration due to their simplicity (without phase change) and the small number of parameters necessary for their description. But the same approach can be applied to any other two-heat-source thermodynamic cycle.

2.5.2.1. *Application to the thermodynamic cycles of ideal gases (Stirling and Ericsson)*

The assumption of a working fluid respecting the ideal gas state equation makes it possible to link the specific heat capacities at constant pressure and constant volume, denoted as, respectively, c_P and c_V to the specific constant of the ideal gas under consideration $r = {}^{R}/_{M}$ and the adiabatic index (or Laplace coefficient) $\gamma = {}^{c_P}/_{c_V}$, according to the following Laplace equations:

$$c_P = r\frac{\gamma}{\gamma-1} \text{ and } c_V = \frac{r}{\gamma-1} \qquad [2.37]$$

Based on this assumption, the Carnot, Stirling and Ericsson cycles can be plotted in a T-s diagram as shown in Figure 2.11.

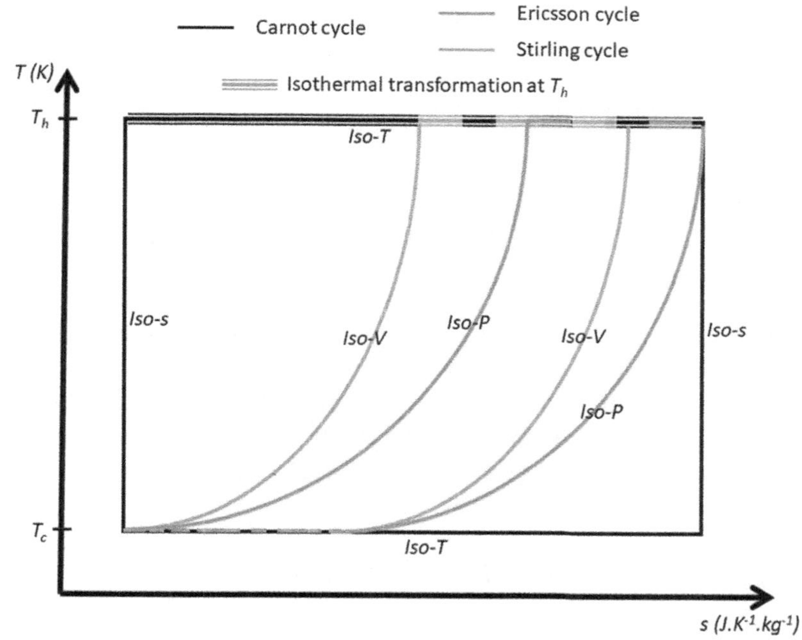

Figure 2.11. *Representation of Carnot, Stirling and Ericsson cycles in a T-s diagram. For a color version of this figure, see www.iste.co.uk/feidt/advances.zip*

These different cycles are characterized by the thermodynamic transformations detailed below, which make it possible to establish the specific energy balances exchanged over a complete cycle, from the ratio of the specific volumes v^* involved in the isothermal transformation at temperature T_h (shown in Figure 2.11), defined as follows:

$$v^* = \frac{v_{q_h}^{initial}}{v_{q_h}^{final}} \qquad [2.38]$$

By definition, the ratio of specific volumes v^* is positive and will have a value lower than 1 in the case of an isothermal expansion at temperature T_h (clockwise cycle) and higher than 1 in the case of isothermal compression at temperature T_h (anti-clockwise cycle).

The specific energy balances exchanged over a complete cycle can thus be written as follows:

– *Carnot cycle* (two isothermal transformations and two isentropic transformations):

$$q_h{}^C = -rT_h \, ln(v^*)$$
$$q_c{}^C = rT_c \, ln(v^*) \quad [2.39]$$
$$w^C = r(T_h - T_c) \, ln(v^*)$$

– *Stirling cycle* (two isothermal transformations and two isochoric transformations):

In order to improve the energy conversion efficiency of this cycle, a heat regenerator can be implemented in order to recover part of the heat released during the exothermic isochoric transformation to be restored during the endothermic isochoric transformation phase. In this general case, the specific energy balances exchanged over a complete cycle are written as follows with a heat regenerator of efficiency ε_r^S:

$$q_h{}^S = (1 - \varepsilon_r^S)c_v(T_h - T_c) - rT_h \, ln(v^*)$$
$$q_c{}^S = -(1 - \varepsilon_r^S)c_v(T_h - T_c) + rT_c ln(v^*) \quad [2.40]$$
$$w^S = r(T_h - T_c) ln(v^*)$$

It can be verified that in the case of an ideal heat regenerator ($\varepsilon_r^S = 1$), the Stirling cycle is equivalent to a Carnot cycle (6–10).

– *Ericsson cycle* (two isothermal transformations and two isobaric transformations):

As in the case of the Stirling cycle, a heat regenerator can be implemented in order to recover part of the heat released during the exothermic isobaric transformation to be restored during the endothermic isobaric transformation phase. In this general case, the specific energy balances exchanged over a complete cycle with a heat regenerator of efficiency ε_r^E are then:

$$\begin{aligned} q_h{}^S &= (1-\varepsilon_r^E)c_p(T_h-T_c) - rT_h\, ln(v^*) \\ q_c{}^S &= -(1-\varepsilon_r^E)c_p(T_h-T_c) + rT_c ln(v^*) \\ w^S &= r(T_h-T_c)ln(v^*) \end{aligned} \qquad [2.41]$$

It can be verified that in the case of an ideal heat regenerator ($\varepsilon_r^E = 1$), the Ericsson cycle is also equivalent to a Carnot cycle (Lhuillier 1992; Pérez 2001; Moran 2008; Borel 2010; Bejan 2016).

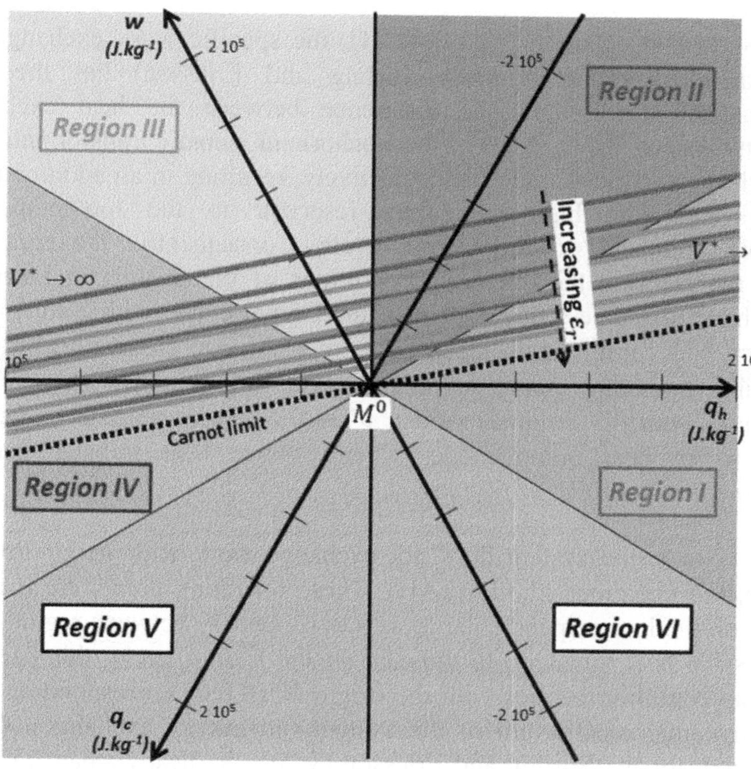

Figure 2.12. *Carnot (black), Stirling (green) and Ericsson cycles of an air ideal gas ($\gamma = 1.4$ and $r = 286.7\, J \cdot kg^{-1} \cdot K^{-1}$) for $\theta = {}^2/_3$ ($T_h = 450\, K$ and $T_c = 300\, K$). Considered heat regenerator efficiencies ε_r varying between 0 to 1 in increments of 0.2. For a color version of this figure, see www.iste.co.uk/feidt/advances.zip*

From these energy balances, the operating points reached for these three cycles with ideal gas as working fluid can be represented in the ternary diagram. For illustration, the operation of these thermodynamic cycles is shown in Figure 2.12, for $T_h = 450$ K and $T_c = 300$ K and for different efficiency values of the heat regenerator ε_r^S and ε_r^E varying between 0 and 1.

In accordance with the previously introduced Carnot limit, the Carnot cycle is represented by a straight line passing through the origin whose angle is $\alpha^C = 1.3807$ rad for $\theta = {}^2/_3$. One can also check that all the possible operating conditions are located in the upper half-plane delimited by the Carnot limit. It should be noted again that the Stirling and Ericsson cycles superimposed with the Carnot cycle when the efficiency of the regenerator is ideal $\varepsilon_r^S = \varepsilon_r^E = 1$.

As shown in equations [2.39]–[2.41], the specific work exchanged for these different cycles (Carnot, Stirling and Ericsson) has the same mathematical expression. The difference between the heat exchanged between these cycles is linked to the isochoric or isobaric transformations of the Stirling and Ericsson cycles, respectively, resulting in an additional heat transfer from the high-temperature reservoir to the low-temperature reservoir, compared to the Carnot cycle. Consequently, the parametric curves representing the operating conditions of the Stirling and Ericsson cycles have the same shape as that of the Carnot cycle, with a shift on the axis $w = 0$, because of this additional term relative to passive heat transfer. Reversible operating conditions can only be approximated ($\alpha \rightarrow \alpha^C$) for theoretical specific volume ratios: $v^* \rightarrow 0$ in heat engine mode and $v^* \rightarrow +\infty$ in heat pump mode, when passive heat transfer becomes negligible.

For $v^* = 1$, the system does not exchange work with its environment ($w = 0$, see equations [2.39]–[2.41]). These operating points are therefore located on the line perpendicular to axis w (boundary between regions I and II – passive heat transfer). In the case of the Carnot cycle, this particular operating condition merges with the origin $M°(0,0,0)$ corresponding to the thermodynamic equilibrium of the system (no energy exchange). On the other hand, for the Stirling and Ericsson cycles, the heat exchanged during isochoric or isobaric transformations result in passive heat transfer from the high temperature heat source to the low-temperature heat sink ($q_h = -q_c$, border between regions I and II), if the heat regenerator is not ideal.

For all the cycles represented, the heat engine mode ($w < 0$, region I) is reached for ratios of specific volumes $v^* < 1$, implying that the cycle is operated clockwise.

Conversely, for $v^* > 1$, the thermodynamic cycles are operated counterclockwise and the system is therefore qualified as a receptor $w > 0$. Different modes of operation (forced heat transfer, heat dissipation and heat pump modes, regions II to IV, respectively) are then possible, depending on the ratio of the specific volumes v^* considered. For low specific volume ratios ($exp\left[-\frac{(1-\varepsilon_r)c}{r}(1-\theta)\right] > v^* > 1$), isochoric or isobaric passive heat transfer (from the high-temperature source to the low-temperature sink) dominate $q_h > 0$ and $q_c < 0$, involving operation in forced heat transfer mode (region II). Heat dissipation mode (region III) is achieved when $exp\left[-\frac{(1-\varepsilon_r)c}{r}\left({}^1/_\theta - 1\right)\right] > v^* > exp\left[-\frac{(1-\varepsilon_r)c}{r}(1-\theta)\right]$. Under these conditions, the isochoric or isobaric passive heat transfer is counterbalanced by the isothermal transformation at the temperature T_h ($q_h < 0$). Finally, high specific volume ratios ($v^* > exp\left[-\frac{(1-\varepsilon_r)c}{r}\left({}^1/_\theta - 1\right)\right]$) are necessary for the Stirling or Ericsson cycles to be able to operate in heat pump mode (region IV) in order to compensate for the heat transfer during the isothermal transformation at the temperature T_c ($q_c > 0$).

Finally, this representation allows us to show that the same operating conditions of a Stirling and Ericsson cycle (superposition of the parametric curves) can be achieved if the following condition on the respective regenerator efficiency of each of the cycles is respected:

$$\frac{1-\varepsilon_r^S}{1-\varepsilon_r^E} = \gamma \qquad [2.42]$$

It can be concluded that the same performance can be achieved for a Stirling cycle with a regenerator with an efficiency of 0.3, as for an Ericsson cycle with a regenerator with an efficiency of 0.5 (in the case of an air ideal gas, with $\gamma = 1.4$).

2.5.3. *General case – endo- and exo-irreversible systems*

Finally, the case of a Stirling cycle of an air ideal gas ($\gamma = 1.4$ and $r = 286.7\ \text{J}\cdot\text{kg}^{-1}\cdot\text{K}^{-1}$) for $T_h = 450$ K and $T_c = 300$ K ($\theta = {}^2/_3$), with a heat regenerator efficiency ε_r^S of 0.8 and thermal resistances between the system and its reservoirs R_h^{th} and R_c^{th} equal to $1\,10^{-3}\,\mathrm{K \cdot kg \cdot J^{-1}}$ is represented in Figure 2.13. This particular case aims to illustrate the graphical analysis of the performance of a real system (endo- and exo-irreversible).

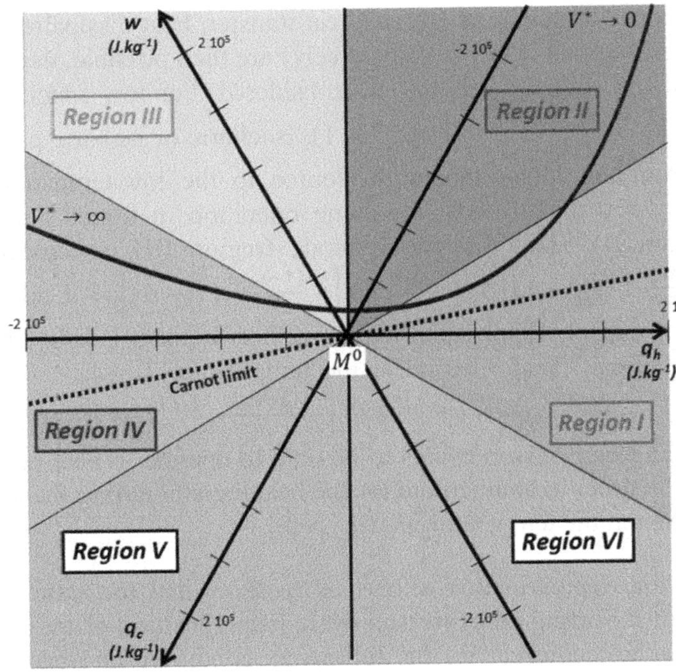

Figure 2.13. *Stirling cycles of an air ideal gas ($\gamma = 1.4$ and $r = 286.7\,J \cdot kg^{-1} \cdot K^{-1}$) for $\theta = {}^2/_3$, ($T_h = 450\,K$ and $T_c = 300\,K$), $\varepsilon_r^S = 0.8$ and $R_h^{th} = R_c^{th} = 1 \times 10^{-3}\,K \cdot kg \cdot J^{-1}$. For a color version of this figure, see www.iste.co.uk/feidt/advances.zip*

The different modes of operation are achieved depending on the ratio of the specific volumes involved during the isothermal transformation at high temperature v^*. The operating conditions at maximum conversion efficiencies (in heat engine and heat pump modes) can be deduced by the two operating points presenting maximum values of angle α. The specific energies exchanged can then be directly read from the graph. In the same way, the operating conditions at maximum powers (for both engine and heat pump modes) can also be deduced graphically, and the corresponding conversion efficiencies via the measurement of the angle α at these points.

In order to support the following discussions on the optimal operation of a thermodynamic cycle, the conversion efficiencies are represented according to the specific useful energy exchanged for both the heat engine and heat pump modes, as shown in Figure 2.14.

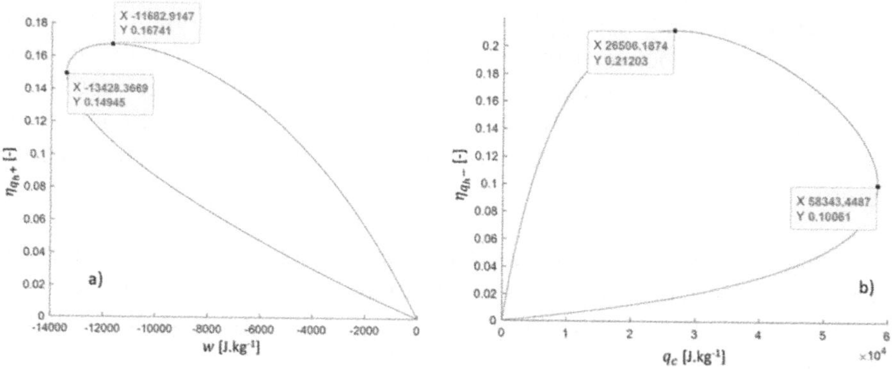

Figure 2.14. *Efficiency as a function of the useful power exchanged in the heat engine (a) and heat pump (b) modes for the Stirling cycle ideal gas ($\gamma = 1.4$ and $r = 286.7\,J \cdot kg^{-1} \cdot K^{-1}$) for $\theta = {}^{2}/_{3}$ ($T_h = 450\,K$ and $T_c = 300\,K$), with $\varepsilon_r^S = 0.8$ and $R_h^{th} = R_c^{th} = 1 \times 10^{-3}\,K \cdot kg \cdot J^{-1}$. For a color version of this figure, see www.iste.co.uk/feidt/advances.zip*

It appears obvious from these representations that the operation conditions of two-heat-source thermodynamic cycles results from a compromise between maximum conversion efficiency (η_{q_h+} maximum in engine mode or η_{q_h-} maximum, equivalent to COP_h and COP_c maximums, in heat pump mode) and a maximum useful energy density (maximum specific work in heat engine mode or maximum specific heat in heat pump mode). This choice can be guided according to the intended application, according to the following general principles:

– if the energy source requested by the system (specific heat at high temperature for engine mode or specific work for heat pump mode) is "free" (available in abundance and/or at low cost), the system can be designed based on the operating point at maximum specific useful energy, in order to recover large quantities of this resource as useful energy, to the detriment of conversion efficiency;

– conversely, if the energy source requested by the system is limited (in quantity or cost), it will then be preferable to size the system for maximum conversion efficiency to the detriment of the useful specific energy produced.

In the vast majority of real applications, it will be a question of seeking a compromise between maximum useful energy and maximum conversion efficiency according to the application context. It will therefore be necessary to size the system between the two extreme cases introduced above. Any other operating point outside this optimal operating range could be improved, either in terms of useful energy density (with equal conversion efficiency), or in terms of conversion efficiency (with equal useful specific energy).

Finally, the final design of the system can be deduced by determining the mass of working fluid (or its flow rate) in order to align the selected operating point, thus setting the useful specific energy, with the targeted useful energy (or the useful power desired, respectively).

2.6. Conclusion and prospects

This original representation is a tool for a simple and unified graphical analysis of the operating conditions of two-heat-source systems. The different achievable operating modes (heat engine, forced heat transfer, heat dissipation and heat pump modes) are thus easily identified and interpreted according to the signs of the specific energy flows exchanged between the system and its environment. This representation, similar to the Raveau diagram, also has the advantage of being based on the standardized axes $(q_{h},; q_c; w)$ related by a rotational symmetry of ${}^{2\pi}/_{3}$ with the origin $M°(0,0,0)$ for center of rotation, resulting from the application of the first law of thermodynamics to two-heat-source systems. It is then possible to significantly deepen the graphical interpretation via the introduction of polar coordinates: the distance r from the operating point considered $M(q_h, q_c, w)$ to the origin $M°(0,0,0)$ is relative to the intensity of the specific energy fluxes exchanged between the system and its environment, and the angle α between the perpendicular to the axis q_c and the operating point considered $M(q_h, q_c, w)$ is directly related to its conversion efficiency. The introduction of an original definition of the conversion efficiency in the case of a high-temperature heat sink, suggested by the symmetry, allowed us to limit the conversion efficiency between 0 and 1 for the potentially reversible operating modes (heat engine and heat pump modes) and at negative values for the dissipative modes (forced heat transfer and heat dissipation modes).

In addition, the application of the second law of thermodynamics makes it possible to reduce the operating range of the system considered, to the upper half-plane delimited by the Carnot limit, represented by a straight line passing through the origin $M°(0,0,0)$, whose slope is set by the reservoir temperatures. Finally, the influence of internal and external sources of irreversibility on the operating conditions of classic two-heat-source cycles could have been highlighted graphically. For all these reasons, this representation proves to be very useful for teaching thermodynamics to students and leads to additional reflections for the study and characterization of the operating conditions of these systems widely used in our societies.

There are many and varied prospects for this work. First, it would be interesting to report on the operating points of the various conversion systems in operation in order to be able to carry out a graphical comparison of the powers exchanged and their respective efficiencies (in engine mode and in heat pump mode). This analysis should make it possible to identify trends, or even recommendations, on the most efficient systems, depending on the cycle implemented, the reservoir temperatures and the targeted applications. Note here that the target objective for these systems may vary depending on the needs relating to the application and accessibility of heat reservoirs: in the case of "free" thermal sources (external air, large-capacity geothermal or aquathermal sources, solar-thermal, etc.), it may be preferable to seek the operating point at maximum power, to the detriment of the system efficiency. Conversely, in the case of limited thermal resources (in time and or space), the search for the operating point with maximum efficiency may be preferred, even if it means reducing the useful power available. In the majority of real applications, it will be necessary to seek a compromise between these operations at maximum useful power and at maximum conversion efficiency.

This representation also made it possible to highlight the link between the deviation from real operation to reversible operation and the corresponding entropy generation by the system (equation [2.31]). This observation deserves to be deepened in order to propose a similar graphical analysis, based on an exergy analysis of the energy flows exchanged by two-heat-source systems, as proposed in Borel (2010).

Finally, the proposed ternary diagram representation should be generalized to all thermodynamic systems, ensuring an energy conversion between two reservoirs of distinct thermochemical potentials based on the analogy between the thermal and thermochemical potentials and fluxes. A first attempt in this direction was made via an analogy with potential energy, as proposed in Figure 2.15.

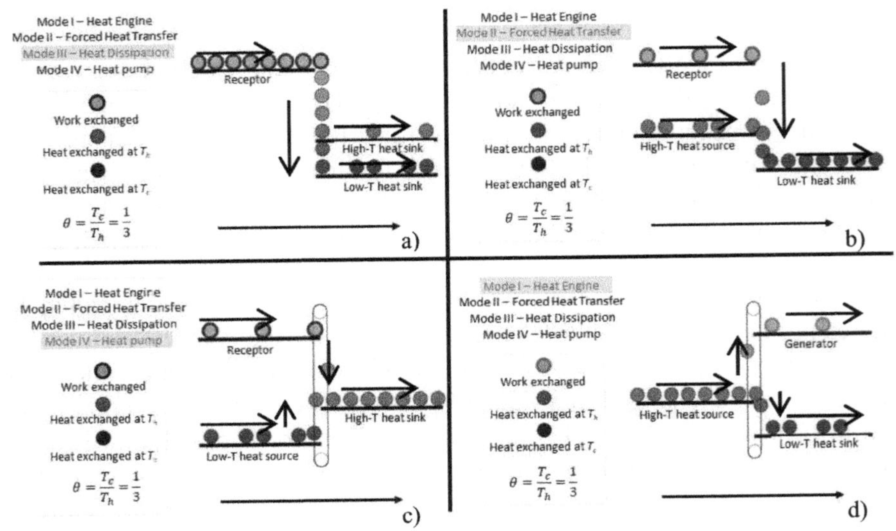

Figure 2.15. *Illustration of the operation of two-heat-source systems by analogy with potential energy. (a) Heat dissipation mode. (b) Forced heat transfer mode. (c) Heat pump mode. (d) Heat engine mode. For a color version of this figure, see www.iste.co.uk/feidt/advances.zip*

In this illustrative diagram, the balls circulate from left to right and the platform level is relative to the exergy potential of the energy flows exchanged by the system with its environment. Thus, the work will always be represented by the platform at the highest exergy level, the high-temperature reservoir by the intermediate level platform (whose height varies according to the thermal potential of the reservoir) and the low-temperature heat reservoir by the lower platform. The different modes observed can thus be described by the circulation of the balls from the upper platforms to the lower platforms (dissipation of energy) and will only be able to reach an upper platform using a pulley system where the energy potential of the rising ball must be provided by the descending ball toward a platform of lower potential (as is the case for the heat engine and heat pump modes).

2.7. References

Alefeld, G. and Radermacher, R. (1993). *Heat Conversion Systems*. CRC Press, Boca Raton, FL.

Bejan, A. (2016). *Advanced Engineering Thermodynamics*, 4th edition. Wiley, Hoboken, NJ.

Borel, L. and Favrat, D. (2010). *Thermodynamics and Energy Systems Analysis: From Energy to Exergy*, Vol. 1. EPFL Press, Lausanne.

Canort, S. (1897). *Reflections on the Motive Power of Heat, and on Machines Fitted to Develop that Power*. John Wiley & Sons, NY.

Chambadal, P. (1957). *Les centrales nucléaires*. Armand Colin, Paris.

Clapeyron, E. (1960). Memoir on the motive power of heat. In *Reflections on the Motive Power of Fire by S. Carnot and Other Papers on the Second Law of Thermodynamics*, Mendoza, E. (ed.). NY Dover Publications, Mineola, NY.

Clausius, R. (1851). On the moving force of heat, and the laws regarding the nature of heat itself which are deducible therefrom. *Phil. Mag.*, 2, 102–119.

Curzon, F.L. and Ahlborn B. (1975). Efficiency of a Carnot engine at maximum power output. *Am. J. Phys.*, 43, 22–24.

Gibbs, J.W. (1873). A method of geometrical representation of the thermodynamic properties of substances by means of surfaces. *Transactions of Connecticut Academy*, 2, 382–404.

Glansdorff, P. and Prigogine, I. (1971). *Structure, stabilité et fluctuations*. Masson et Cie, Paris.

Haase, R. (1968). *Thermodynamics of Irreversible Processes*. Addison-Wesley, Reading, MA; Menlo Park, CA; London.

Joule, J.P. (1850). On the mechanical equivalent of heat. *Phil. Trans. R. Soc.*, 140, 61–82.

Lhuillier, C. and Rous, J. (1995). *Introduction à la thermodynamique*, 3rd edition. Dunod, Paris.

Mollier, R. (1904). *Neue Diagramme zur Technischen Wärmelehre*. VDI-Z, Berlin.

Moran, M., Shapiro, H., Boettner, D., Bailey, M. (2014). *Fundamentals of Engineering Thermodynamics*, 8th edition. John Wiley & Sons, Hoboken, NJ.

Novikov, I.I. (1958). The efficiency of atomic power stations (a review). *J. Nucl. Energy II*, 7, 125–128.

Pérez, J.-P. (2001). *Thermodynamique : fondements et applications*, Vol. 1, 3rd edition. Dunod, Paris.

Prigogine, I. (1968). *Introduction à la thermodynamique des processus irréversibles*. Dunod, Paris.

Ramousse, J. (2016). Représentation graphique des modes opératoires des systèmes dithermes. In *SFT 2016, Proceedings of congrès annuel de la Société Française de Thermique*, Toulouse, 31 May–3 June 2016.

Ramousse, J. (2018). Ternary diagram of bithermal systems. *International Journal of Thermodynamics*, 21(1), 27–36.

Raveau (2021). Diagramme de Raveau [Online]. Available at: https://fr.wikipedia.org/wiki/Diagramme_de_Raveau [Accessed February 15, 2021].

Zahra, A.-M. and Mathieu, J.-C. (1989). *Thermodynamique des processus irréversibles*. Techniques de l'Ingénieur, Saint-Denis.

3

Thermodynamics with Finite Speed

Stoian PETRESCU and Monica COSTEA

Department of Engineering Thermodynamics, Heat Engines, Thermal and Refrigeration Equipment, Politehnica University of Bucharest, Romania

3.1. Introduction

This chapter presents the essence of thermodynamics with finite speed (TFS), as a new branch of irreversible thermodynamics for engineers, and the direct method derived from it, for the optimization of *heat engines* (engines, refrigerators and heat pumps, conventional or solar) and electrochemical devices (fuel cells, electrolyzers, batteries) from the point of view of *power and efficiency (or COP)*, when the internal and external irreversibilities of the processes are taken into account.

All heat engines, as well as all electrochemical devices that operate with a *finite speed w*, respectively a *finite current density i*, operate irreversibly. Because of this, for all these devices *a method to optimize them* is necessary, so as to work as close as possible to the optimum efficiency, in order to have minimum irreversibilities (generation of minimum entropy) or maximum power or efficiency.

This method, called the direct method, is based on the new expression of the first law of thermodynamics for finite speed processes, which actually combines the first law with the second law in a single equation that can be

Advances in Thermodynamics and Circular Thermoeconomics,
coordinated by Michel FEIDT and Antonio VALERO-CAPILLA. © ISTE Ltd 2023.

integrated for any irreversible process/cycle in order to directly obtain analytical expressions of efficiency and power. These expressions can be used for the optimization of any irreversible cycle.

The main stages in the development of TFS and the direct method will be presented: the first developments and fundamental concepts; expression of the first law for irreversible processes in simple and complex systems; invention and application of the direct method; development of the Stirling engine performance calculation model and validation; application of TFS for the Carnot, Otto, diesel, semi-diesel engine cycles; and new areas of research for the extension of TFS.

Some conclusions will close the presentation by reviewing the results of the approach based on TFS and the direct method.

3.2. First developments of TFS (1964–1974)

3.2.1. *The origin of TFS*

The beginning of the development of TFS is marked by the article published in 1964 by Stoian Petrescu and his thesis supervisor at the time, Lazar Stoicescu (Stoicescu and Petrescu 1964a), followed by work for his doctoral thesis (Petrescu 1969a). Thus, inspired by books (Macke 1963; Sommerfeld 1967) and Einstein's theory of relativity (Einstein 1955; Petrescu 1971a), the authors wrote for the first time a new equation of the first law of thermodynamics for processes at finite speed in simple closed systems (piston cylinder) in the following form:

$$dU = \delta Q - P_{m,i} \cdot \left[1 \pm \frac{aw}{c}\right] \cdot dV \qquad [3.1]$$

where $P_{m,i}$ is the instantaneous mean gas pressure, w is the mean piston speed, a is the coefficient which depends on the nature of the gas and $c = \sqrt{3RT_{m,i}}$ – average molecule speed (Clausius expression), with $T_{m,i}$, instantaneous average temperature and R, specific gas constant.

Equation [3.1] marks an important moment, because for the first time the cause of irreversibility has been contained in a fundamental equation of thermodynamics, namely the mathematical expression of the first law, written especially for irreversible processes, and which causes the *finite piston speed w* and the *finite speed of molecules c* to appear.

The analogy with Einstein's theory of relativity consists of the fact that, similar to the dependence of the speed of light on the interactions of electromagnetic waves, here, in the cylinder-piston assembly, the waves generated by the movement of the piston with the finite speed w depend on the average speed of the molecules c. This is precisely the reason why the interaction between the piston and the gas can be written in the form of the ratio w/c. Thus, this ratio accounting for the irreversibility due to the finite speed of the piston led to equation [3.1], using four different approaches, two kinetic-molecular and two phenomenological (section 3.2.3.3).

One of these approaches was based on the famous theory of Onsager and Prigogine, *linear irreversible phenomenological thermodynamics*, using the concept of entropy generation. Thus, despite the name given initially to equation [3.1] – the first law for finite speed processes – it, actually, results from a combination of the first law and the second law of thermodynamics, due to the presence of the *irreversibility factor (1 +/– aw/c)*. This parenthesis differentiates between the first law for reversible processes (when $w \rightarrow 0$, which leads to the classical expression of the first law for reversible processes) and the expression when w is non-zero (finite value).

Both a faster speed w and a slower speed c (lower temperature) will produce more irreversibilities. The term aw/c, which is representative of the cause of irreversibility, is differentiated in equation [3.1] using the new sign (+/–), with (+) for compression and with (–) for expansion. This change of sign differentiates the two processes, in particular by the effect of the irreversibility on the mechanical work, because during the compression, more work is necessary compared to the process of reversible compression, and on the contrary, during expansion, less work is provided compared to the reversible process. These aspects of irreversibility were known at the time (in 1964), but only *qualitatively* and figured in almost every important book on thermodynamics. What was not known at the time was a *quantitative expression for irreversible work*, which is the second term on the right-hand side of equation [3.1]:

$$\delta W_{irr} = P_{m,i}\left(1 \pm \frac{aw}{c}\right) dV \qquad [3.2]$$

with $P_{m,i}$ as instantaneous mean pressure in the system.

This pressure is part of the set of new concepts introduced by TFS for the study of irreversible processes, which will be discussed in the following section.

3.2.2. *Fundamental concepts*

In irreversible TFS, where non-equilibrium conditions prevail, new concepts and conditions are needed to describe the state of the system and the processes the system undergoes. This analysis becomes more complex than when considering classical reversible processes. However, TFS more closely models real systems, both qualitatively and quantitatively.

The cylinder-piston device shown in Figure 3.1 shows the simplest closed system in which an irreversible finite speed process can take place. In this case, the finite speed interactions between the system and the environment are only thermal, δQ_{irr}, and mechanical, δW_{irr}. By the correlation of the pressures inside the cylinder, the figure presents the new concepts about the pressure variation during the expansion process with finite speed.

3.2.2.1. *Instantaneous pressure distribution in the system (gas)*

The curve indicated by $P_{gaz,i}$ is described in the new type of P-V diagram introduced in irreversible TFS, the instantaneous gas pressure in the cylinder (system), at a certain moment of the piston's motion at speed w. We do not know exactly what this distribution looks like, so the simplest approximation we can consider is the linear one, as shown in Figure 3.1. In fact, this inclined line connects the pressure exerted by the gas on the piston P_p and the cylinder head P_c. and each of its points accounts for the instantaneous gas pressure at a location (point or section) in the cylinder.

3.2.2.2. *Mean instantaneous gas pressure*

Since it is impossible to use the instantaneous gas pressure at every point in the system and at every instant during a process, the concept of a mean instantaneous pressure $P_{m,i}$ is necessary.

The dotted horizontal line indicates the uniform distribution that would occur in the system if the piston stops and the finite expansion time of the system is reached.

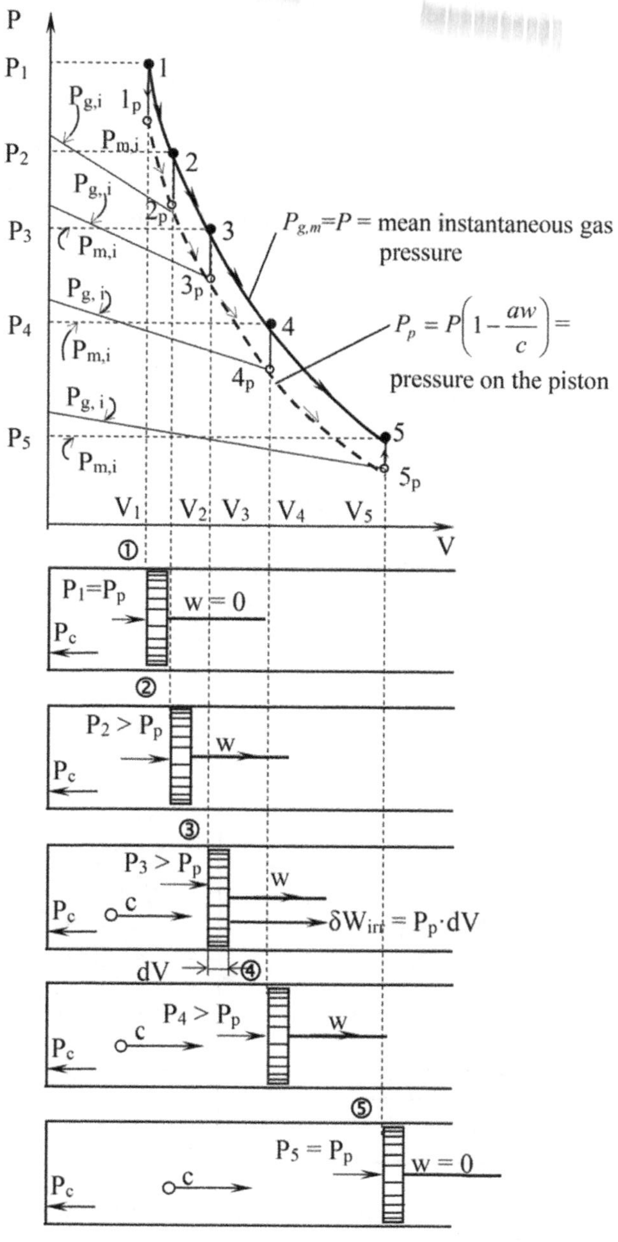

Figure 3.1. *Irreversible expansion with finite speed in a simple thermodynamic system*

This pressure $P_{m,i}$ will be used to describe the processes in the system by replacing the equilibrium pressure P from reversible thermodynamics at an equilibrium. The main difference between what happens in finite-speed processes and reversible equilibrium processes, where the speed approaches zero, is the need to use two pressures instead of one to correctly describe finite-speed processes. Indeed, in TFS, the gas pressure of the system $P_{m,i}$ and the pressure on the piston P_p are different, as shown by the two curves in Figure 3.1.

3.2.2.3. *Pressure on the piston*

The pressure on the piston differs from the gas pressure at any point in the system, as shown by the pressure distribution curves in Figures 3.1 and 3.2. Thus, in an expansion process, the pressure on the piston P_p will be lower than the gas pressure at any other point in the system, reaching a maximum at the cylinder head P_c, as shown in Figure 3.1.

On the contrary, during a compression process, the pressure on the piston P_p will be greater than the pressure at any other point in the system. The gas pressure in the system reaches a minimum at the cylinder head P_c.

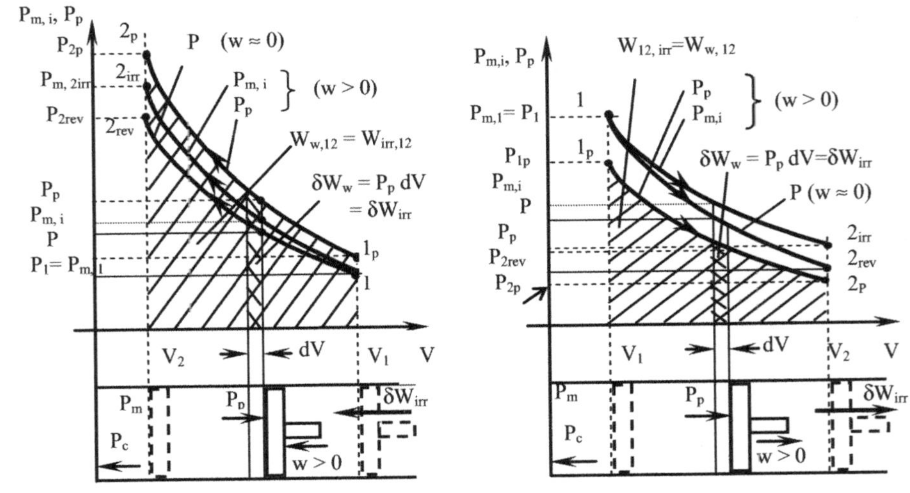

Figure 3.2. *Comparison of reversible and irreversible finite-speed processes in a simple thermodynamic system: (a) compression; (b) expansion*

A better perspective on the important differences in process analysis between reversible equilibrium thermodynamics (RET) and the new irreversible TFS is shown by the coupling of the curves in Figures 3.2(a) and 3.2(b). Note that the irreversible processes have been plotted using a simplification of the two parameters P_p and $P_{m,i}$.

The two curves previously discussed, $1_p - 2_p$ and $1–2_{irr}$, are represented in each figure, as a consequence of the irreversibility due to the finite speed and by comparison, the curve of the reversible process, $1 - 2_{rev}$. In both irreversible processes, the gas pressure curve $1–2_{irr}$ is greater than that corresponding to the reversible process $1 - 2_{rev}$, determining higher gas state parameters at the end of the compression or expansion process.

As for the mechanical work, taking into account equation [3.2], there is a clear difference between the work corresponding to the irreversible process compared to the reversible one (see the areas below the curves $1_p–2_p$ in the two diagrams), both in adiabatic compression and expansion. As a result, more mechanical work is needed in compression and less is provided in expansion compared to the corresponding reversible process.

3.2.2.4. *Mean instantaneous temperature and mean instantaneous specific internal energy*

In TFS, the temperature varies from point to point at every instant and also varies with time at every point during the processes. Therefore, it is necessary to introduce the concepts of a mean instantaneous temperature $T_{m,i}$ and an instantaneous specific internal energy $u_{m,i}$, similar to the introduction of the concept of mean instantaneous pressure $P_{m,i}$.

The mean instantaneous temperature is conceptually and quantitatively different from the equilibrium temperature T, which is uniformly distributed in the system.

3.2.3. *The first law of thermodynamics for finite-speed processes in simple systems*

For a simple closed system, in which the effects of piston-cylinder friction and internal gas viscosity are neglected, the mathematical expression of the first law for irreversible processes with finite speed is written as:

$$dU_{irr} = \delta Q_{irr} - \delta W_{irr} \quad [3.3]$$

This expression seems identical to that for reversible processes (which corresponds to w → 0), but each term of [3.3] implies different parameters compared to those of equilibrium used for reversible processes.

3.2.3.1. *Internal energy in TFS*

The variation in the internal energy is based on the new concepts specific to TFS, in particular the mean instantaneous temperature $T_{m,i}$ and the mean instantaneous specific internal energy $u_{m,i}$:

$$dU_{irr} = m \cdot du_{m,i} = m \cdot c_v \cdot dT_{m,i} \qquad [3.4]$$

with c_v as the specific heat at constant volume of the gas (J/kgK).

Equation [3.4] seems to be almost the same as the expression for the variation in internal energy in a reversible process, but it is not. It is important to note the meaning of $T_{m,i}$ and the means of determining it, both theoretically and by actual measurements. These calculations are more complex in TFS than in RET.

3.2.3.2. *Heat transfer in TFS*

The heat transfer accounts for the irreversibilities due to the existence of a temperature gradient between the gas and the external heat source. Since a *temperature distribution* $T_{m,i}$ exists in the system, when heat passes from an external fluid to the fluid inside a system through a solid wall, the term δQ_{irr} is expressed as follows:

$$\delta Q_{irr} = \dot{Q}_{irr} \cdot d\tau = U_g \cdot A \cdot \left(T_{m,i} - T_S\right) \cdot d\tau \qquad [3.5]$$

where U_g is the overall heat transfer coefficient (W/m^{2}K); A is the heat transfer surface area (m^2); T_S is the temperature of the heat source (K).

Here, we could draw a parallel with the expression for heat in a reversible process, but the presence of the mean instantaneous temperature in [3.5] makes the difference.

3.2.3.3. *Mechanical work in TFS*

The irreversible mechanical work in TFS involves the pressure on the piston (see Figure 3.1), which is not the same as the mean instantaneous gas pressure in the cylinder, $P_{m,i}$. Therefore, one of the most important problems in the development of TFS is to find an expression for P_p as a function of $P_{m,i}$ and various other system parameters, such as piston speed w, the mean instantaneous temperature $T_{m,i}$.

It took 7 years (1964–1971) for Petrescu and Stoicescu to develop four finite velocity gas-piston interaction models in closed systems (Stoicescu and Petrescu 1964a, 1964b; Petrescu 1969a, 1969b, 1971b), and 9 years (1964–1973) to develop a finite velocity gas–gas interaction model in open systems (Petrescu 1973). We present here the results obtained from the four models concerning closed systems:

– The simple kinetic-molecular model (Stoicescu and Petrescu 1964a; Petrescu 1969a) leads to the expression:

$$\delta W_{irr} = P_{m,i} \cdot \left[1 \pm \frac{aw}{c} + \frac{bw^2}{c^2} \pm \ldots\right] \cdot dV = P_p \cdot dV \quad [3.6]$$

with $a = 2$, $b = 5$ and $c = \sqrt{3RT_{m,i}}$, where R [J/kgK] is the gas constant.

– The advanced kinetic-molecular model (Petrescu 1969a, 1971b), based on the Maxwell–Boltzmann distribution (Boltzmann 1898) and the concept of finite-time expansion in the system, provides:

$$\delta W_{irr} = P_{m,i} \cdot \left[1 \pm 2.764\frac{w}{c} + 3\frac{w^2}{c^2} \pm 1.283\frac{w3}{c^3} + \ldots\right] \cdot dV \quad [3.7]$$

– The phenomenological analysis based on the propagation of the pressure waves generated by the finite speed of the piston and the speed of sound in the gas (Stoicescu and Petrescu 1964a; Petrescu 1969a) leads to:

$$W_{irr} = P_{m,i} \cdot \left[1 \pm \frac{aw}{c}\right] \cdot dV \text{ or } \delta W_{irr} = P_{m,i} \cdot \left[1 \pm \gamma\left(\frac{w}{c_s}\right)\right] \cdot dV \quad [3.8]$$

where $c_s = \sqrt{\gamma RT_{m,i}}$, the speed of sound in gas, and $a = \sqrt{3\gamma}$, with $\gamma = c_p/c_v$.

– The approach based on the second law (Petrescu 1969b), using the linear irreversible phenomenological thermodynamics of Onsager and Prigogine, which led to the development of a finite-speed gas-piston interaction and the expression:

$$\delta W_{irr} = P_{m,i} \cdot [1 \pm K_I \cdot w] \cdot dV \quad [3.9]$$

with K_I a constant that depends on the properties and the mean instantaneous temperature of the gas.

This approach also considers the second law of thermodynamics with the entropy generation determined by the finite speed of the piston, which is present in the parenthesis in expression [3.9]. Exactly this will introduce the irreversibilities into all the equations that characterize the processes in heat engines and that will be deduced from the first law, combined with equation [3.9].

This combination of the first and second law of thermodynamics will provide the tool to *quantitatively discuss irreversible processes* in heat engines and to develop *calculation schemes that allow validation.*

Due to the fact that only equation [3.8] provides an expression for the constant a, this one will be used for numerical applications:

$$W_{irr} = P_{m,i} \cdot \left[1 \pm \frac{aw}{c}\right] \cdot dV = P_p \cdot dV \qquad [3.10]$$

3.2.3.4. *The fundamental equation of irreversible TFS*

Replacing expressions [3.5] and [3.10] in equation [3.1] leads to an initial first law expression for finite-speed processes in simple systems in the form:

$$dU_{irr} = U_g A\left(T_{m,i} - T_S\right) \cdot d\tau - \left(1 \pm \frac{aw}{c}\right) \cdot P_{m,i} \cdot dV \qquad [3.11]$$

In this form, equation [3.11] is valid regardless of the working fluid. In the case of an ideal gas, the variation in the irreversible internal energy (equation [3.4]) is substituted with that of the mean instantaneous pressure $P_{m,i}$ expressed from the state equation $P_{m,i} = \frac{mRT_{m,i}}{V}$. Added to this is the duration of the process, expressed as a function of the piston speed w and the sign convention for the heat transfer. Altogether, this provides:

$$mc_v dT_{m,i} = \pm U_g A\left(T_{m,i} - T_S\right) \cdot \frac{dV}{wA_p} - \left(1 \pm \frac{aw}{c}\right) \cdot \frac{mRT_{m,i}}{V} \cdot dV \qquad [3.12]$$

In the dimensionless form, this results in an expression with two variables, a dimensionless temperature and a dimensionless volume:

$$\frac{c_v \cdot dT_{m,i}}{RT_{m,i}} = \left[\pm \frac{U_g AV}{wA_p mR}\left(1 - \frac{T_S}{T_{m,i}}\right) - \left(1 \pm \frac{aw}{c}\right)\right] \cdot \frac{dV}{V} \qquad [3.13]$$

Thus, *the TFS fundamental equation* was obtained and known as the *first law of thermodynamics for irreversible processes with finite piston speed* (Petrescu 1969a, 1969b, 1971, 1973, 1974; Stoicescu and Petrescu 1964a, 1964b, 1965a, 1965b, 1965c).

Unfortunately, the variables of this expression cannot be separated, so a direct integration is not possible. However, the equation can be solved by:

– numerical integration, although very precise, it does not provide an analytical equation for irreversible processes at finite speed;

– analytical integration for special cases;

– preliminary estimation of some average parameters during the process, which will be used to improve either the analytical integration, or the numerical solution by successive iterations.

3.2.3.5. *Analytical study of the five irreversible transformations based on the first law for processes at constant or variable finite speed*

The beginning of the development of finite speed thermodynamics is based on equation [3.1] and all these new concepts needed to describe irreversible processes. This equation introduces irreversibility, due to the finite speed of the piston, which determines a non-uniform distribution of the pressure inside the system.

Considering the terms of equation [3.11], its specific integration was done for adiabatic transformation, which seemed the most accessible due to the term of heat transfer being equal to zero. Thus, the following temperature–volume equation of an irreversible adiabatic process with finite speed was obtained:

$$T_1 V_1^{(\gamma-1)} \cdot \alpha_1 = T_2 V_2^{(\gamma-1)} \cdot \alpha_2 \qquad [3.14]$$

Then, using the state equation, the other two analytical equations were deduced in the form:

$$P_1 V_1^{\gamma} \cdot \alpha_1 = P_2 V_2^{\gamma} \cdot \alpha_2 \qquad [3.15]$$

$$T_1 P_2^{\frac{\gamma-1}{\gamma}} \cdot \alpha_1^{\frac{1}{\gamma}} = T_2 P_1^{\frac{\gamma-1}{\gamma}} \cdot \alpha_2^{\frac{1}{\gamma}} \qquad [3.16]$$

with:

$$\alpha_1 = \left(1 \pm \frac{aw}{c_1}\right)^2, \quad \alpha_2 = \left(1 \pm \frac{aw}{c_2}\right)^2 \qquad [3.17]$$

where $a = \sqrt{3\gamma}, \quad c_1 = \sqrt{3RT_1}, \quad c_2 = \sqrt{3RT_2}$.

It is remarkable that these three equations have expressions similar to the equations of the reversible adiabatic transformation, but with the addition of the *irreversibility factor* α, which is dependent on the speed w.

Other irreversibility factors resulted from the integration of equation [3.1] for the other irreversible thermodynamic transformations (isobaric, isochoric, isothermal and polytropic) with the piston at *constant speed* (Stoicescu and Petrescu 1964b). They are highlighted in Table 3.1, which gives an overview of the five irreversible transformations, whose cause of irreversibility is the finite speed of the piston.

The same analytical approach has been developed for thermodynamic transformations with the piston (Stoicescu and Petrescu 1965a; Petrescu 1969a) at *variable speed (sinusoidal)*. Taking into account the quasi-sinusoidal movement of the piston in real engines introduced a correction factor of 1.24, which multiplied the term *aw/c*.

It is important to note here that an expression very close to equation [3.14] was discovered by Petrescu much later on (in 1968), during his Fulbright visiting researcher internship at Rensselaer Polytechnic Institute, Troy, USA, in an article published by Paul (1937).

The German physicist studied the first finite speed process in a cylinder-piston system and obtained the first equation of a finite speed irreversible adiabatic process in the form:

$$TV^{(\gamma-1)\left(1 \pm 2.60067 \cdot \frac{w}{c} \pm 3 \cdot \frac{w^2}{c^2} \pm 3.6743 \cdot \frac{w^3}{c^3}\right)} = cst \qquad [3.18]$$

Process	Mass	Equation	ΔU, ΔH	W_{12}	Q_{12}
Isobaric $P_2 = P_1$	$m = \frac{P_1 V_1}{R T_1}$	$\frac{T_1}{V_1} = \frac{T_2}{V_2}$	$\Delta U = m c_v \Delta T$ $\Delta H = m c_p \Delta T$	$W_{12} = P_1(V_2 - V_1) \cdot \boldsymbol{\sigma}$	$Q_{12} = \Delta H = m\, c_p \,\Delta T \cdot \boldsymbol{\sigma_Q}$
Isometric $V_2 = V_1$	$m = \frac{P_1 V_1}{R T_1}$	$\frac{P_1}{T_1} = \frac{P_2}{T_2}$	$\Delta U = m c_v \Delta T$ $\Delta H = m c_p \Delta T$	$W_{12} = 0$	$Q_{12} = \Delta U = m\, c_v \,\Delta T$
Isothermal $T_2 = T_1$	$m = \frac{P_1 V_1}{R T_1}$	$P_1 V_1 = P_2 V_2$	$\Delta U = 0$ $\Delta H = 0$	$W_{12} = P_1 V_1 \ln \frac{V_2}{V_1} \boldsymbol{\beta}$	$W_{12} = P_1 V_1 \ln \frac{V_2}{V_1} \boldsymbol{\beta}$
Adiabatic $\delta Q_{rev} = 0$ γ – ratio of specific heats or adiabatic exponent	$m = \frac{P_1 V_1}{R T_1}$ R – ideal gas constant	$P_1 V_1^{\gamma} = P_2 V_2^{\gamma} \cdot \frac{\boldsymbol{\alpha_2}}{\boldsymbol{\alpha_1}}$ $T_1 V_1^{\gamma-1} = T_2 V_2^{\gamma-1} \cdot \frac{\boldsymbol{\alpha_2}}{\boldsymbol{\alpha_1}}$ $\frac{T_1}{P_1^{\frac{\gamma-1}{\gamma}}} = \frac{T_2}{P_2^{\frac{\gamma-1}{2}}} \left(\frac{\boldsymbol{\alpha_2}}{\boldsymbol{\alpha_1}}\right)^{1/\gamma}$	$\Delta U = m c_v \Delta T$ $\Delta H = m c_p \Delta T$	$W_{12} = \frac{P_1 V_1 - P_2 V_2}{k - 1}$	$Q_{12} = 0$
Polytropic n – polytropic exponent	$m = \frac{P_1 V_1}{R T_1}$	$p_1 V_1^n = p_2 V_2^n$ $T_1 V_1^{n-1} = T_2 V_2^{n-1}$ $T_1 / p_1^{\frac{n-1}{n}} = T_2 / p_2^{\frac{n-1}{n}}$	$\Delta U = m c_v \Delta T$ $\Delta H = m c_p \Delta T$	$W_{12} = \frac{P_1 V_1 - P_2 V_2}{n - 1}$	$Q_{12} = \Delta U + W_{12}$
Corrections due to finite speed	$\boldsymbol{\alpha_1} = \left(\mathbf{1} \pm \frac{\boldsymbol{aw}}{\boldsymbol{c_1}}\right)^2$; $c_1 = \sqrt{3RT_1}; a = \sqrt{3\gamma}$ $\boldsymbol{\alpha_2} = \left(\mathbf{1} \pm \frac{\boldsymbol{aw}}{\boldsymbol{c_2}}\right)^2$; $c_2 = \sqrt{3RT_2}$		$\boldsymbol{\beta} = \left(1 \pm \frac{aw}{c_T}\right)$ ← T = ct irreversible $\boldsymbol{\sigma} = \left(1 \pm \frac{aw}{c_{12,m}}\right)$; $c_{12,m} = \frac{c_1 + c_2}{2}$ ← P = ct irreversible $\boldsymbol{\sigma_Q} = \left(1 \pm \frac{aw}{c_{12,m}} \cdot \frac{W_{12,rev}}{\Delta H}\right)$ ← P = ct irreversible - heat		

Table 3.1. *Table of irreversible thermodynamic processes*

Most likely, Paul's interest at this time (early WWII) was to more accurately calculate the speed of the "piston" (a projectile) in a gun, compared to approximate calculations, based on the classical reversible adiabatic process, which considers the piston speed w tending to zero.

The similarity between equations [3.14] and [3.18] is truly remarkable:

– same notations are used: w for the speed of the piston and c for the average speed of the molecules;

– use of the sign (+/–) in order to differentiate the irreversible processes of compression and expansion.

The only difference is the position of the "irreversibility factor", which is placed in the exponent in equation [3.18], while it multiplies the terms in equation [3.14].

However, there is an essential difference between the two approaches in that Paul studied a single irreversible process while the five irreversible processes shown in Table 3.1 were studied based on the same fundamental equation [3.1], which in fact combines the first and second law of thermodynamics.

The verification of this equation was made by an experiment carried out at the Polytechnic Institute of Romania (Stoicescu and Petrescu 1965c) and continued by Petrescu with the help of Prof. Van Ness at Rensselaer Polytechnic Institute (Petrescu 1974).

3.2.4. *First approach to Beau de Rochas and Otto's finite speed irreversible cycle*

For the first time in the history of TFS, the irreversible finite-speed Beau de Rochas and Otto cycle approach had been developed analytically using the equations in Table 3.1. It showed the efficiency in the sense of the first law of thermodynamics of this irreversible cycle in the form (Stoicescu and Petrescu 1965b):

$$\eta_{I,BdR\&O,irr,w} = \left(1 - \frac{1}{\epsilon^{\gamma-1}}\right) \cdot k_{corr,w} = \eta_{I,BdR\&O,rev} \cdot k_{corr,w} \quad [3.19]$$

where $k_{corr,w}$ is the correction factor taking into account the irreversibility due to the finite speed of the piston, which we will name the irreversibility factor:

$$k_{corr,w} = \left[1 - \frac{2\,a \cdot w}{c_1} \cdot \frac{1}{\varepsilon^{\frac{\gamma-1}{2}}\left(\varepsilon^{\frac{\gamma-1}{2}}+1\right)\left(\lambda^{\frac{1}{2}}-1\right)}\right] \quad [3.20]$$

where ε is the compression ratio λ is the isochoric combustion pressure ratio.

The influence of the mean piston speed w_m on the reversible efficiency correction factor is shown in Figure 3.3. There is decrease in the value of this factor with the increase in the finite speed of the piston. However, for usual values of the speed w in heat engines, the factor $k_{corr,w}$ determines a moderate reduction in the efficiency of the irreversible cycle, which will no longer be the case when other irreversibilities (throttling and friction) will be taken into account (see sections 3.3.1.1 and 3.3.1.2).

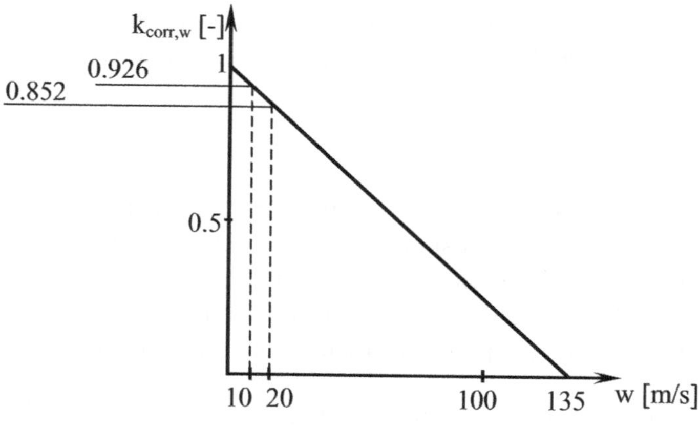

Figure 3.3. *Variation in the irreversibility factor present in the expression of the efficiency of the irreversible cycle of Beau de Rochas and Otto, as a function of piston speed*

Figure 3.3 also shows the theoretical maximum speed corresponding to the cancellation of the correction factor $k_{corr,w}$, which results analytically from expression [3.20] as per the following equation (Stoicescu and Petrescu 1965b):

$$w_{max} = \frac{c_1}{2\,a} \cdot \varepsilon^{\frac{\gamma-1}{2}} \left(\varepsilon^{\frac{\gamma-1}{2}} + 1\right)\left(\lambda^{\frac{1}{2}} - 1\right) \qquad [3.21]$$

Also note that in all works after 1965, where we studied the efficiency of a real irreversible cycle within the framework of thermodynamics at finite speed, we used the same analytical model of efficiency (for engine cycles) or COP (for reverse cycles – refrigerator and heat pump) as in expression [3.19], as the product between the efficiency of the reversible cycle, $\eta_{I,rev}$ and the irreversibility factor as a function of the speed w and other constructive and functional parameters of the engine.

3.3. Developments of TFS during the period 1990–2007

3.3.1. *First law of thermodynamics for finite speed processes in complex closed systems*

The 1990s saw new developments (Petrescu 1991; Petrescu et al. 1992; Petrescu and Harman 1994) that supplemented the fundamental TFS equation [3.1] for irreversible processes in real heat engines with terms evaluating two other causes of irreversibility, namely throttling and piston-cylinder friction.

3.3.1.1. *Effect of pressure loss due to throttling*

Complex closed systems contain regenerative heat exchangers (or throttling device included) that simple closed systems do not have and which are responsible for an internal irreversibility that is added to the one generated by the finite speed of the piston (non-uniformity of pressure in the system or dual effect of gas pressure and pressure on the piston).

A closed complex system is shown in Figure 3.4 (Petrescu et al. 1992). It is assumed that an ideal gas flows through a regenerative heat exchanger (made of a network of thin tubes or porous material) as a result of the movement of the piston in the cylinder. The regenerative heat exchanger divides the cylinder into two compartments, I and II.

During a finite speed expansion process, the gas passes from compartment I to compartment II, which results in the loss of pressure through the heat exchanger and as a result, an instantaneous pressure distribution, $p_{i,gas}$. This pressure distribution is represented (Figure 3.4) for positions 1 and 2 of the piston.

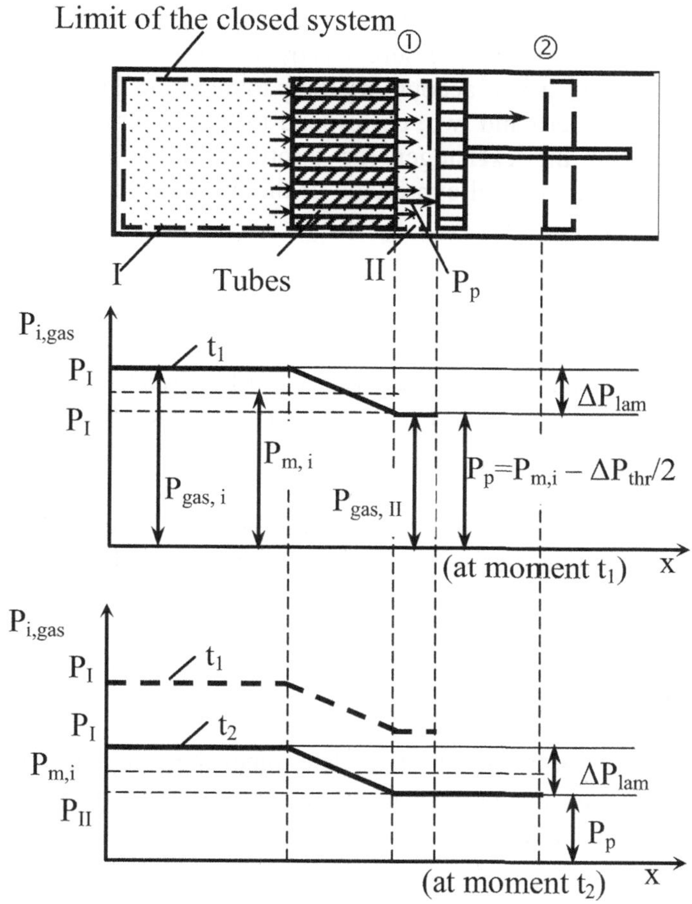

Figure 3.4. *Instantaneous distribution of gas pressure in a cylinder with a regenerative heat exchanger during a finite speed expansion process*

The pressure on the piston, *Pp*, decreases as a result of pressure loss due to throttling as the gas passes through the regenerator. In terms of mean instantaneous pressure, we write:

$$P_p = P_{m,i} - \frac{\Delta P_{lam}}{2} = P_{m,i}\left(1 - \frac{\Delta P_{lam}}{2P_{m,i}}\right) \qquad [3.22]$$

During a finite speed compression process, the variation in the instantaneous pressure changes the direction, and the pressure on the piston will increase (the sign "–" is replaced with "+" in the parentheses).

In addition, the position of the exchanger in the cylinder leads to a different contribution of the pressure loss by throttling to the pressure on the piston, which is taken into account by the parameter b. The extreme values of this parameter, zero and two, correspond to the maximum and minimum distance between the exchanger (or throttling device) and the piston, respectively. Thus, the final expression of the pressure on the piston results as:

$$P_p = P_{m,i} - \frac{\Delta P_{throt}}{2} = P_{m,i}\left(1 \pm \frac{b \cdot \Delta P_{throt}}{2P_{m,i}}\right) \qquad [3.23]$$

with sign (+) for compression and the sign (–) for expansion, and $0 < b < 2$ (Petrescu 1991; Petrescu et al. 1992, Petrescu and Costea 2011).

3.3.1.2. *Effect of pressure loss by friction on piston pressure*

Friction between piston and cylinder is an important source of irreversibility in finite speed processes in closed systems. Figure 3.5 presents such a system in which an ideal gas undergoes a process of adiabatic compression with friction. The first law of thermodynamics requires that the increase in the internal energy of the gas be equal to the decrease in energy of its surroundings.

The movement of the piston on dx determines the variation of the volume dV and a reduction in the mechanical work due to friction ($\delta W_f = F_f\,dx$), which increases the internal energy of the system (cylinder and piston).

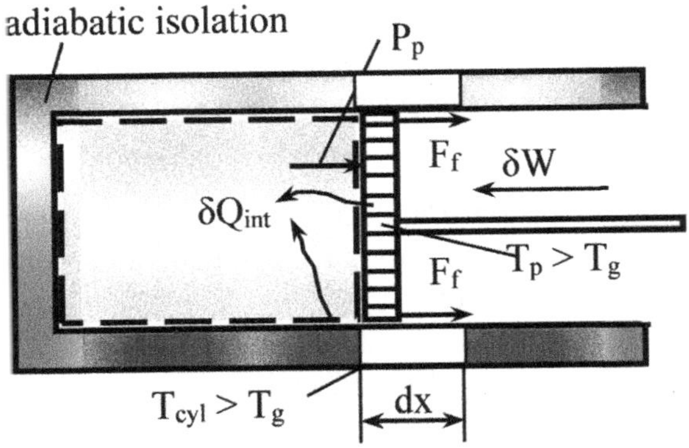

Figure 3.5. *Adiabatic compression with friction in a closed system*

Thus, the temperature of the cylinder will increase up to T_{cyl}, while that of the piston will increase up to T_p, both temperatures being above the gas temperature ($T_{cyl} > T_g$ and $T_p > T_g$). Due to these temperature differences, heat transfer should occur between the cylinder and the gas. Since the cylinder is adiabatic, the heat received by the gas is:

$$\delta Q_{int,cpr} = f\left|\delta W_f\right| > 0 \qquad [3.24]$$

with $0 < f < 1$, the coefficient that shows how much of the heat produced by friction remains in the gas.

The expression of the first law for an adiabatic compression process with friction is written as:

$$dU_{g,cpr} = -f \cdot \Delta P_f \mathrm{d}V - P_{m,i,cpr} \cdot \mathrm{d}V = -P_{m,i,cpr}\left(1 + f\frac{\Delta P_f}{P_{m,i,cpr}}\right)\mathrm{d}V \qquad [3.25]$$

Similar reasoning in the case of adiabatic expansion with friction will lead to a similar expression, but with the sign "–" in parentheses:

$$\mathrm{d}U_{g,exp} = -P_{m,i,det}\left(1 - f\frac{\Delta P_f}{P_{m,i,exp}}\right)\mathrm{d}V \qquad [3.26]$$

Returning to the general case of a closed system (piston cylinder), which includes both mechanical and thermal interactions with the environment and the effects of friction, we can write the expression of the first law in the form:

$$\mathrm{d}U_{irr} = \delta Q_{irr} - P_{m,i}\left(1 \pm \frac{f \cdot \Delta P_f}{P_{m,i}}\right)dV \qquad [3.27]$$

with + for compression and – for expansion.

3.3.1.3. *Expression of the first law of thermodynamics for finite speed processes in complex closed systems*

A general expression of the first law of thermodynamics for finite speed processes in complex closed systems can be written by superimposing the causes of internal irreversibility during a finite speed compression or expansion process. The causes of internal irreversibility were as follows: (1) the finite speed of the piston, (2) the throttling of the gas and (3) the mechanical friction between piston and cylinder. It results in as follows:

$$dU = \delta Q - P_{m,i}\left(1 \pm \frac{aw}{c} \pm \frac{b \cdot \Delta P_{throt}}{2P_{m,i}} \pm \frac{f \cdot \Delta P_f}{P_{m,i}}\right) dV \qquad [3.28]$$

The expressions of the pressure losses by throttling and by friction will be detailed in the following sections for each heat engine, but always as a function of the piston speed. By expressing all the losses according to the same variable helps to unify their analysis, just according to the most important parameter, the speed of the piston *w*. This was the key to success in optimizing this speed for maximum efficiency or for maximum power of Stirling engines first, and later for any irreversible cycle, such as the Carnot, Otto, Diesel, Brayton and Otto-Stirling combined cycle (Petrescu and Costea 2011; Petrescu et al. 2015a).

To the expression of the first law for finite speed processes, we always associate that of the mechanical work:

$$\delta W_{irr} = P_{m,i}\left(1 \pm \frac{aw}{c} \pm \frac{b \cdot \Delta P_{throt}}{2P_{m,i}} \pm \frac{\Delta P_f}{P_{m,i}}\right) dV \qquad [3.29]$$

which is not quite identical to that of the right side of equation [3.24], which we would be tempted to use, except for $f = 1$. This is because the irreversibility introduced by friction determines the integral demand/loss of corresponding work in equation [3.25].

3.3.2. *The direct method of TFS*

The method used for the analysis of finite speed irreversible processes involves:

– integrating the new mathematical expression of the first law of thermodynamics for finite speed processes, in each type of irreversible process;

– deducing the equations of the five thermodynamic processes (isobaric, isochoric, isothermal, adiabatic and polytropic) and the associated energy exchanges, *taking into account the internal irreversibility*;

– *analytically* assessing the efficiencies of heat engines and electrochemical devices, which has not been done by any other branch of irreversible thermodynamics (linear irreversible phenomenological thermodynamics of Onsager-Prigogine, entropy method, exergy method, finite time thermodynamics), despite their success.

This technique is called the direct method. The fact that the analytical expressions of the efficiencies contain the speed of the process, the geometrical parameters (volumes, rate of compression, etc.) as well as the properties of the fluid (specific heats, molar mass, adiabatic exponent) makes possible the optimization of these parameters or the study of their influence on the targeted efficiencies. This is the power of the direct method: it can be used in a systematic way, step by step, similar to that of reversible thermodynamics, but now taking into account internal irreversibilities on fundamental bases (equation [3.25]).

3.4. Main achievements of TFS and the direct method

3.4.1. *Modeling and optimization of the Stirling engine cycle*

The Stirling engine cycle consists of two isothermal transformations (1-2, compression and 3-4, expansion) and two isometric transformations (2-3, heating and 4-1, cooling), as shown in the diagrams *p-V* and *T-S* in Figure 3.6.

In addition to being an external combustion engine, the constructive particularity that differentiates it from other heat engines is the presence of the regenerator – a heat exchanger made up of a porous metal matrix capable of storing the heat released by the gas during its passage in one direction (4-1) and to transfer this heat to the gas at the passage in the opposite direction (2-3).

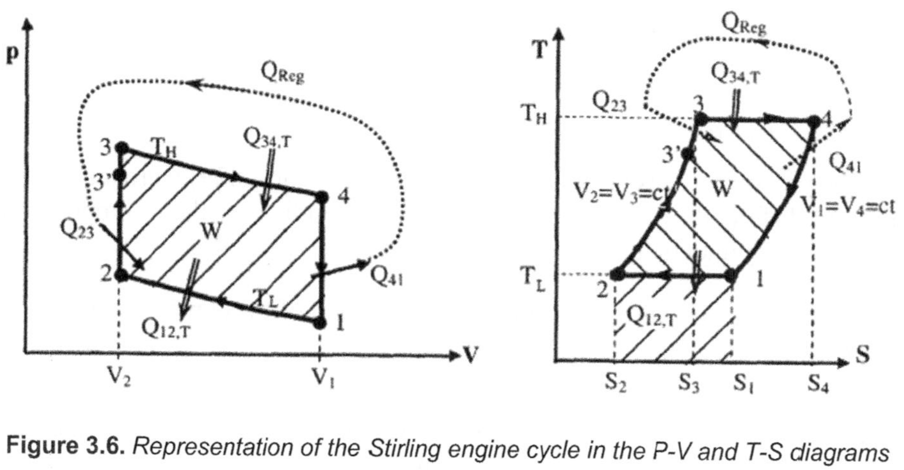

Figure 3.6. *Representation of the Stirling engine cycle in the P-V and T-S diagrams*

The integral transfer of heat in the regenerator (called perfect regeneration) assuming either an infinite transfer surface area, or an infinite time of contact between gas and the porous matrix, is associated with the ideal cycle, while the real process (imperfect regeneration) is accompanied by heat losses. So, instead of having the equality $Q_{23} = |Q_{41}|$, the heat actually regenerated will be $Q_{23'} < Q_{23}$. The efficiency of the regenerator is expressed as:

$$\sigma_R = \frac{Q_{23'}}{Q_{23}} = \frac{Q_{23'}}{|Q_{41}|} < 1 \qquad [3.30]$$

or more useful in the analytical approach would be *the heat loss coefficient in the regenerator* defined as:

$$X = 1 - \sigma_R \qquad [3.31]$$

The consequences of imperfect regeneration will be analytically assessed in the engine efficiency calculation scheme in the next section.

3.4.1.1. *Calculating the finite speed Stirling engine cycle efficiency*

This technique is based on the first law of thermodynamics for finite speed processes in complex systems (Petrescu et al. 1992; Petrescu and Harman 1994) and the direct method for closed systems. The merit of this scheme lies in its ability to express the efficiency of the engine, as well as all the terms involved in it (imperfect regeneration, pressure losses due to friction, finite speed and throttling processes in the regenerator) as a function of the mean speed of the Stirling engine piston and other geometric characteristics (piston stroke, diameters, compression ratio, number of regenerator screens).

3.4.1.1.1. Expression of pressure loss

The pressure losses present in equations [3.28] and [3.29] were taken into account differently for each type of engine, depending on their composition (number of cylinders, presence of valves or regenerator, etc.).

For the Stirling engine, the analytical approach of these pressure losses is based on:

– Studies by Organ (1992), who synthesized numerous empirical measures of ΔP_{throt} as a function of the Reynolds (*Re*) number of the gas passing through the regenerator screens. Our adapted formula that best describes the empirical results of Organ and gives the piston speed dependence is:

$$\Delta P_{throt} = \frac{15}{\gamma}\rho_R\frac{w_R^2}{2}N \text{ [bar]} \qquad [3.32]$$

where ρ_R is the density of the gas in the regenerator [kg/m^3], w_R is the gas velocity in the regenerator [m/s], and N is the number of regenerator screens.

– Transformation of Heywood's empirical equations (Heywood 1988) for frictional pressure losses, ΔP_f, in internal combustion engines (ICE), expressed as a function of rotational velocity N_r:

$$\Delta P_f = 0.97 + 0.15\frac{N_r}{1000} \text{ [bar]} \qquad [3.33]$$

in an expression as a function of piston speed:

$$\Delta P_f = 0.94 + 0.45\frac{w}{100 \cdot S} \text{ [bar]} \qquad [3.34]$$

with N_r as the rotation speed (rot/min), w as the piston speed (m/s) and S as the piston stroke (m).

The expression [3.33] is based on experimental measurements on ICEs and has been correlated empirically. Its use for Stirling engines is justified by the tendency to design and build (technologically) these engines in the same way as ICEs (with similar parts and similar technologies).

3.4.1.1.2. Analytical approach to efficiency

Efficiency of the Stirling engine in the sense of the first law of thermodynamics is expressed as a product of the Carnot cycle efficiency and a second law efficiency (Stoicescu and Petrescu 1964a, 1964b), as suggested by Bejan (1988), and in a similar way to the approach of the first cycle studied using TFS (Stoicescu and Petrescu 1965b):

$$\eta_{MS} = \eta_{CC} \cdot \eta_{II,irr} = \left(1 - \frac{T_{L,g}}{T_{H,g}}\right) \cdot \eta_{II,irr} \qquad [3.35]$$

where the second law of efficiency is expressed by:

$$\eta_{II,irr} = \eta_{II,X} \cdot \eta_{II,\Sigma\Delta P_i} \qquad [3.36]$$

The internal irreversibilities highlighted by the expression [3.36] are *imperfect regeneration* and *pressure losses* (due to finite speed, throttling and friction).

Imperfect regeneration is a major loss in Stirling engines. This loss is expressed using a regenerative heat loss coefficient, X, a concept introduced by Michel Feidt (1987) that inspired us to model efficiency $\eta_{II,X}$. An expression of the second law of efficiency, which takes into account the effect of the heat losses in the regenerator expressed as a function of the speed of the piston, as well as many other parameters (geometric, properties of the gas) has been derived (Florea 1999; Petrescu et al. 2002a) as follows:

$$\eta_{II,X} = \frac{1}{1+\frac{X \cdot c_v}{R \ln \varepsilon}\left(1-\frac{T_{L,g}}{T_{H,g}}\right)} \qquad [3.37]$$

with $T_{L,g}$ as the gas temperature of the low-temperature isotherm of the cycle, $T_{H,g}$ as the temperature of the gas of the high-temperature isotherm of the cycle, ε as the compression ratio.

The heat loss coefficient in the regenerator (X) has been the subject of our previous studies (Florea 1999; Petrescu et al. 2002a; Petrescu and Costea 2011). This is the term that includes all losses due to incomplete heat transfer in the regenerator. It obviously depends on a large number of parameters and variables. Among these are the piston speed w, the dimensions of the cylinder (diameter D_C and stroke S), the dimensions of the regenerator (diameter D_R and length L), the properties of the solid porous material in the regenerator (mass m_R, density ρ_R, specific heat c_R), the dimensions and number of screens (d, b and N), gas properties (m_g, ρ_g, c_p, V, R) and operating conditions ($\varepsilon = V_1/V_2$, $\tau = T_{H,g}/T_{L,g}$, T_m, P_m). A relation for X including the effect of these parameters was evaluated using the first law and heat transfer analysis. This analysis of heat losses led to differential equations that were integrated.

The integration of the equations was based on two complementary approaches:

– an overall analysis of the system (configuration, number of regenerators, type of porous matrix), which gives relatively unfavorable results, in particular a large X_1;

– a linear temperature distribution in the regenerator and gas matrix, which leads to relatively favorable results, namely smaller X_2.

The resulting expressions for X_1 and X_2 are:

$$X_1 = \frac{1+2M+e^{-B}}{2(1+M)}, \qquad X_2 = \frac{M+e^{-B}}{1+M} \qquad [3.38]$$

with

$$M = \frac{m_g c_{v,g}}{m_R c_R}, \qquad B = (1+M)\frac{h\ A_R}{m_g c_{v,g}} \cdot \frac{S}{w} \qquad [3.39]$$

$$h = \frac{0.395\ (4P_m/RT_L)\ w_g^{0,424} \cdot c_P(T_m) \cdot v(T_m)^{0.576}}{(1+\tau)\left[1-\frac{\pi}{4[(b/d)+1]}\right] D_R^{0,576} \cdot Pr^{2/3}} \qquad [3.40]$$

where A_R is the heat transfer surface area of the regenerator screens (m^2), M and B are non-dimensional, h is the convective heat transfer coefficient in a porous medium (Incropera and DeWitt 1996), P_m is the mean gas pressure in the engine (Pa), T_m is the mean gas temperature in the engine (K), $\nu(T_m)$ is the kinematic viscosity of the gas at the mean temperature T_m (m^2/s), d is the diameter of the metal wires on the screens in the regenerator (m), b is the distance between the metal wires on the screens in the regenerator (m) and Pr is the Prandtl number (-).

The calculated values of X_1 and X_2 have been found to predict with good accuracy the values of the heat loss coefficient in the regenerator, X, determined from experimental data available in the literature (Allen and Tomazic 1987; Farell et al. 1988; Geng 1987; Fujii 1990; Stine and Diver 1994) using the following equation:

$$X = yX_1 + (1-y)X_2 \qquad [3.41]$$

where an *adjustment coefficient* is introduced, whose value was calculated at 0.72.

The second efficiency of expression [3.36] accounts for pressure losses in the engine due to the finite speed of the processes, throttling and friction in the regenerator. This analytical approach is based on the calculation of the losses of mechanical work on the processes of the cycle in two similar hypotheses compared to those introduced by X_1 and X_2, but defined here by the *pessimistic and optimistic method.* Thus, the pessimistic method considers mean pressure losses over the four processes of the cycle, while the optimistic one associates them strictly to the specific process (losses by throttling – only in the processes in the regenerator; losses by finite velocity and friction – only in compression and expansion processes).

The performance predicted by the equations of the two methods was compared with the experimental efficiency of a variety of top-performing Stirling engines (Allen and Tomazic 1987; Geng 1987; Farell et al. 1988; Fujii 1990; Stine and Diver 1994) worldwide (the United States, Japan, Sweden). Actual data from the Stirling engine was combined with the above analysis, and a single equation for the efficiency due to the second law of thermodynamics, which takes pressure losses into account, was proposed in the following form:

$$\eta_{II,\Sigma\Delta P_i} = 1 - \frac{\frac{w}{w_{SL}}\gamma(1+\sqrt{\tau})\, ln\,\varepsilon + 5\cdot N\left(\frac{w_R}{w_{SL}}\right)^2 + \frac{3(0.94+0.045\,w)}{4P_1}10^5}{\tau\cdot\eta'\cdot ln\,\varepsilon} \quad [3.42]$$

with w_{SL} as the velocity of the gas at low temperature in the cycle and:

$$\eta' = \left(1-\sqrt{\frac{T_L}{T_{H,S}}}\right)\cdot\left[1+\frac{X\left(1-\sqrt{T_L/T_{H,S}}\right)}{(\gamma-1)\,ln\,\varepsilon}\right]^{-1} \quad [3.43]$$

The mechanical power of the engine introduces the second *adjustment coefficient* of the calculation, z, such that:

$$\dot{W}_{MS,irr} = \eta_{MS}\cdot\dot{Q}_H = \eta_{MS}\cdot z\cdot mRT_{H,g}\cdot\frac{w}{2S}\cdot ln\,\varepsilon \quad [3.44]$$

This adjustment coefficient shows that whatever the amount of heat available at the hot source, the engine can take only a part of it, due to the finite speed ($w > 0$) of the transfer process, responsible for the external irreversibility. The expression [3.44] gives an accurate value for the power of 12 Stirling engines (Allen and Tomazic 1987; Farell et al. 1988; Geng 1987; Fujii 1990; Stine and Diver 1994), when the adjustment coefficient (z) is 0.8 (Petrescu et al. 2002a).

This analytical approach embodied in the Stirling engine efficiency calculation provides insight into the processes involved as it uses the first law of thermodynamics for finite speed processes in conjunction with the direct method. The method simulates the operation of real Stirling engines while intuitively suggesting the mechanisms that generate irreversibilities. The analysis also includes the effect of heat transfer losses in regeneration processes.

Finally, the analysis is used in conjunction with experimental data from a few operational Stirling engines to arrive at a calculation capable of predicting the efficiency of any Stirling engine of known components and configurations.

3.4.1.2. *Validation of the calculation and the direct method*

The results provided by this analysis, which we will call *calculation and optimization of Stirling engines*, are in very good agreement with the experimental efficiencies of 12 different operational Stirling engines over a power range going from an economical regime to that of maximum power. This provides a solid verification that this analysis, based on the direct method, can provide a good estimate of the actual performance of the Stirling engine, in particular with regard to the efficiency and the power delivered. The results reported in Table 3.2 and illustrated in Figure 3.7 are proof of this.

The success of the validation of the direct method and of the Stirling engine calculation and optimization scheme is important, because in comparison with the NASA Code, only two adjustment parameters were used (which is much less compared to the nine parameters used in the NASA code for a single free-piston Stirling engine). This high degree of adaptability of the calculation scheme to different types of Stirling engines has allowed us to extend it to solar sets with the Stirling engine, where it has also been validated (see example in Figure 3.8) (Stanciu 2007; Petrescu and Costea 2011).

This figure is particularly interesting because it also shows the concordance of the calculated and experimental values for the efficiency of the components of the solar set, in particular, that of the receiver η_{Rec} of the engine η_{SE} and overall η_{Tot}, the latter being the only one to present a maximum as a function of the temperature of the gas at the hot end of the engine.

The high degree of correlation between the analytical and operational data illustrated in Figures 3.7 and 3.8 and Tables 3.2 and 3.3 indicates that the analysis based on the direct method can accurately estimate the performance of Stirling engines under all conditions in terms of power and efficiency. This capability has value in the optimized design of new Stirling engines and in estimating the power and efficiency of a particular Stirling engine.

In addition, this calculation deserves special attention for all those interested in understanding and applying the TFS direct method and it is also useful for refrigerators and Stirling heat pumps.

Stirling engine	Experimental power [kW]	Calculated power [kW]	Experimental efficiency [%]	Calculated efficiency [%]
NS-03M, regime 1 (econ.)	2.03	2.18	35.9	33.92
NS-03M, regime 2 ($\dot{W}_{max}$)	3.81	4.19	31	32.97
NS-03T, regime 1 (econ.)	3.08	3.14	32.6	31.9
NS-03T, regime 2 ($\dot{W}_{max}$)	4.14	4.45	30.3	30.9
NS-30A regime 1 (econ.)	23.20	29.45	37.5	35.7
NS-30A, regime 2 ($\dot{W}_{max}$)	30.40	33.82	33	33.6
NS-30S regime 1 (econ.)	30.90	33.78	37.2	36.6
NS-30S, regime 2 ($\dot{W}_{max}$)	45.60	45.62	35.2	35.26
STM4-120	25	26.36	40	40.1
V-160	9	8.82	30	30.8
4-95 MKII	25	28.40	29.4	28.9
GPU-3	3.96	4.16	12.7	12.6
MP1002 CA	0.20	0.19	15.6	15.3
4-275	50	48.61	42	41.19
Free-piston Stirling engine	9	9.16	33	33.10

Table 3.2. *The power and efficiency calculated analytically by the direct method compared to the real performance of several Stirling engines (Florea 1999; Petrescu et al. 2002a). For a color version of this table, see www.iste.co.uk/feidt/advances.zip*

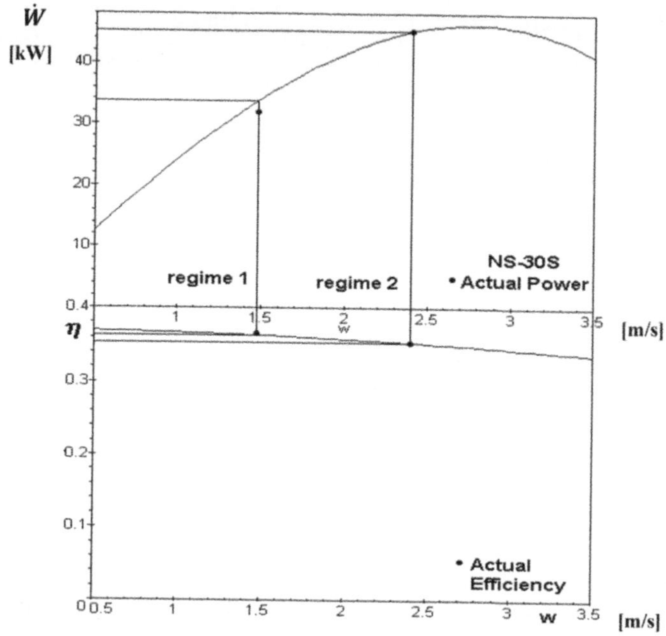

Figure 3.7. *Comparison of analytical results with experimental efficiency of the Japanese Stirling engine NS-30S (Florea 1999; Petrescu et al. 2002a)*

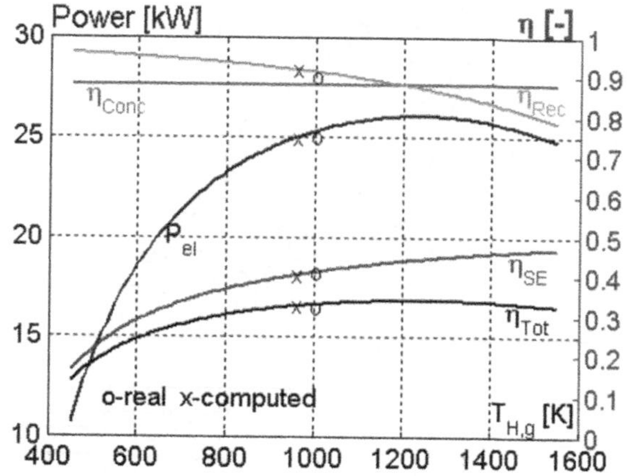

Figure 3.8. *Calculated performances (lines) (Petrescu et al. 2010a; Petrescu and Costea 2011) compared with experimental data (Stine and Diver 1994) of the solar parabolic dish-Stirling engine assembly MDAC (McDonnell Douglas). For a color version of this figure, see www.iste.co.uk/feidt/advances.zip*

Solar set	Receiver efficiency [%]		Engine efficiency [%]		Total efficiency [%]		Electric power [kW]	
	exper.[1]	calc.[2]	exper.	calc.	exper.	calc.	exper.	calc.
Vanguard	90	92	41	41	32.84	33.26	25	25.4
MDAC	90	92	38–42	40	30–33	32.66	25	24.93
SBP 17	80	78	42	40	26.44	24.4	50	48.74

[1] experimental values

[2] calculated values

Table 3.3. *The calculated and experimental performances of solar systems with Stirling engines (Stanciu 2007; Petrescu et al. 2010a; Costea 2011). For a color version of this table, see www.iste.co.uk/feidt/advances.zip*

3.4.2. *Optimization of the Carnot engine irreversible cycle*

The optimization model developed for the Carnot engine cycle (closed system) is also based on the direct method and the first law of thermodynamics for finite speed processes.

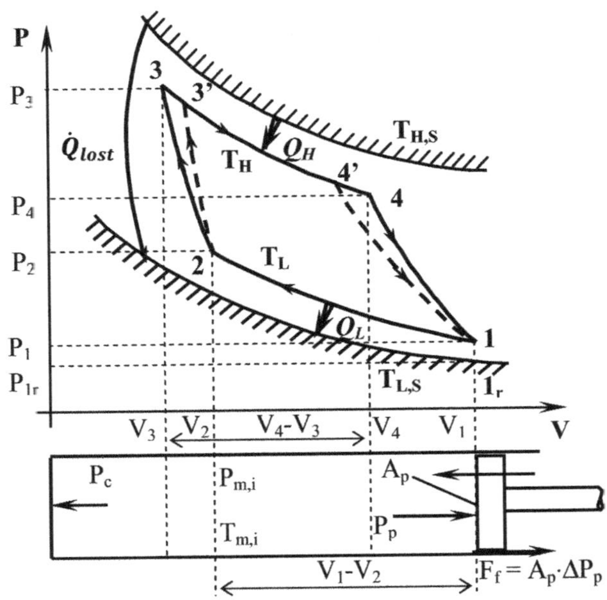

Figure 3.9. *Cycle of the finite speed Carnot engine shown in the P-V diagram (Petrescu et al. 2002; Petrescu and Costea 2011)*

The Carnot engine cycle including the irreversibilities of the heat transfer between the gas in the heat engine and its heat reservoirs, the heat losses between the reservoirs and the internal dissipations of the cycled fluid, is illustrated in Figure 3.9 (Petrescu et al. 2002b; Petrescu and Costea 2011; Costea et al. 2021). The working fluid in the system is alternately in contact with a hot reservoir at the constant temperature $T_{H,S}$ and to a cold reservoir at constant temperature $T_{L,S}$ and its temperatures are, respectively, T_H and T_L. The heat losses between the temperatures of both heat sources through the engine are considered by the flux $\dot{Q}_{lost}$. Moreover, the irreversible adiabatic processes are illustrated by the curves 2-3' and 4'-1.

The mathematical expression of the first law of thermodynamics for finite speed processes in a closed system [3.28] will keep only two terms here as follows:

$$dU = \delta Q - p_{av,i}\left(1 \pm \frac{aw}{c} \pm \frac{f \cdot \Delta p_f}{p_{av,i}}\right) dV \qquad [3.45]$$

and irreversible mechanical work too:

$$\delta W_{irr} = P_{av,i}\left(1 \pm \frac{aw}{c} \pm \frac{\Delta P_f}{P_{av,i}}\right) dV \qquad [3.46]$$

where the pressure losses by friction [3.33], expressed as a function of the piston speed, become in this case:

$$\Delta P_f = (0.97 + 0.045w)/N \qquad [3.47]$$

with N being considered equal to 80 in the example calculation that led to the results shown in Figure 3.10 (Petrescu et al. 2002b; Petrescu and Costea 2011; Costea et al. 2021).

The integration of [3.45] on the isothermal processes of the Carnot cycle leads to expressions that highlight the effect of irreversibilities on the heat exchanges of the cycled fluid with the source Q_H and sink, Q_L:

$$Q_{H,w} = z_H' \cdot mRT_H ln\,\varepsilon \quad ; \quad |Q_{L,w}| = z_L' \cdot mRT_L ln\,\varepsilon \qquad [3.48]$$

where $\varepsilon = \frac{V_1}{V_2} = \frac{V_4}{V_3}$ is the compression ratio.

The irreversibility coefficients dependent on the finite speed of the processes, which appear in the relations of [3.48], show that whatever the heat available at the source, the cycled fluid can only receive a limited quantity of heat, while it will yield a greater quantity to the well compared to the case of the reversible cycle:

$$z_H' = \left(1 - \frac{aw}{\sqrt{3RT_H}} - \frac{f \cdot \Delta p_f}{p_{m,34}}\right) \quad ; \quad z_L' = \left(1 + \frac{aw}{\sqrt{3RT_L}} + \frac{f \cdot \Delta p_f}{p_{m,12}}\right) \qquad [3.49]$$

Similarly, the integration of [3.46] makes it possible to determine the irreversible work produced/consumed during the isothermal processes of the cycle:

$$W_{H,w} = z_H \cdot mRT_H ln\, \varepsilon \quad ; \quad |W_{L,w}| = z_L \cdot mRT_L\, ln\, \varepsilon \qquad [3.50]$$

with the corresponding irreversibility coefficients:

$$z_H = \left(1 - \frac{aw}{\sqrt{3RT_H}} - \frac{\Delta P_f}{P_{m,34}}\right) \quad ; \quad z_L = \left(1 + \frac{aw}{\sqrt{3RT_L}} + \frac{\Delta P_f}{P_{m,12}}\right) \qquad [3.51]$$

which, by the absence of the coefficient $f < 1$, increase the effect of the two causes of irreversibility on the mechanical work compared to that on the heat.

The work in the cycle results from [3.50] as:

$$W_{cycle,w} = W_{H,w} - |W_{L,w}| = mR(z_H T_H - z_L T_L) \cdot ln\, \varepsilon \qquad [3.52]$$

With cycle time determined by speed w and the surface area of the piston, A_p such that:

$$\tau_{cycle} = \frac{2(V_1 - V_3)}{wA_p} \qquad [3.53]$$

results in the expression of the supplied power:

$$P_{\Delta T,w,\dot{Q}lost} = \frac{W_{cycle,w}}{\tau_{cycle}} \qquad [3.54]$$

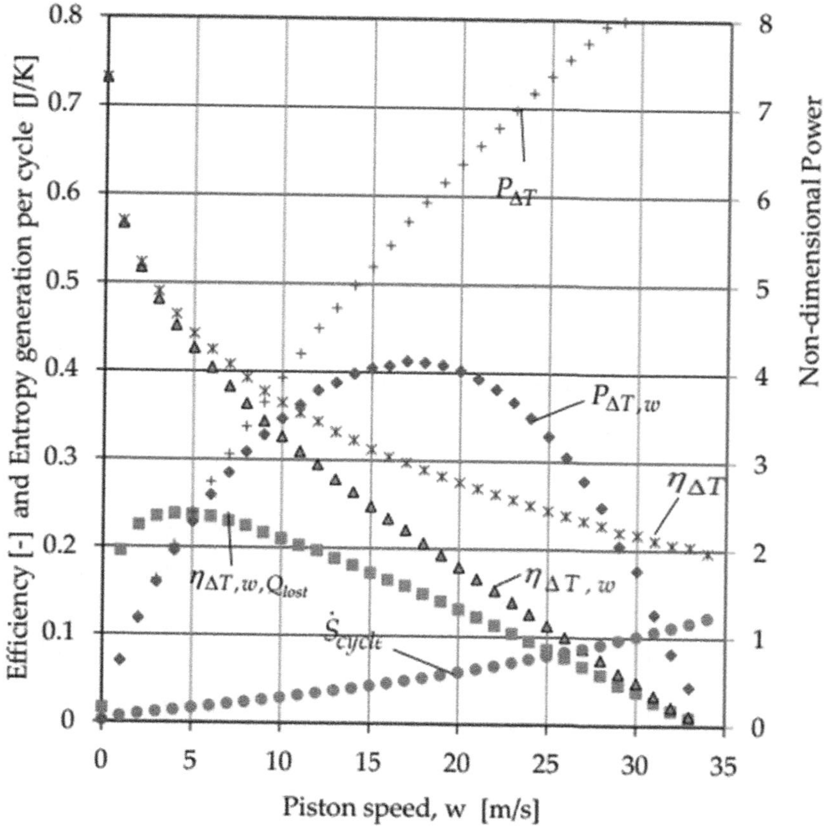

Figure 3.10. *Effect of irreversibilities on the power, efficiency and generation of entropy over time of the Carnot cycle at finite speed. For a color version of this figure, see ww.iste.co.uk/feidt/advances.zip*

The efficiency of the Carnot cycle with internal and external irreversibilities is given by:

$$\eta_{\Delta T,w,Q_{lost}} = 1 - \frac{|Q_{L,w}|}{Q_{H,w}} = 1 - \frac{T_L}{T_H} \cdot \frac{z_L'}{z_H'} \qquad [3.55]$$

The analytical approach also allows the calculation of the entropy generation of the cycle and in time:

$$\Delta S_{cycle} = \frac{Q_{H,w}}{T_H} + \frac{Q_{L,w}}{T_L} = mR\, ln\, \varepsilon \cdot (z_H' - z_L')\ ;\ \dot{S}_{cycle} = \dot{S}_{gen} = \frac{\Delta S_{cycle}}{\tau_{cycle}} \qquad [3.56]$$

An example of the optimization results is illustrated by the curves in Figure 3.10. There are different optimal values of the piston speed corresponding to the maximum power and efficiency, respectively. Also, the maximum power and the efficiency decrease, just like the optimal speed for the maximum, if irreversibilities are added successively into the analysis.

3.4.3. *Effect of irreversibilities on the efficiency of the Beau de Rochas–Otto cycle at finite speed*

One of the studies based on the TFS direct method was conducted on the irreversible Beau de Rochas-Otto cycle (Petrescu et al. 2008; Petrescu and Costea 2011).

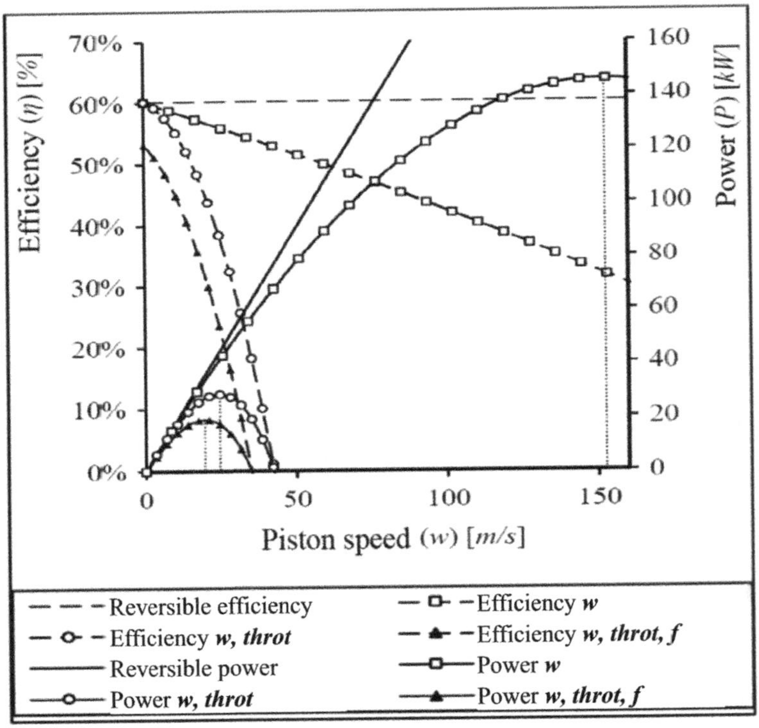

Figure 3.11. *The efficiency and power delivered by the Beau de Rochas-Otto engine cycle as a function of piston speed (Petrescu et al. 2008)*

The optimization results are illustrated in Figure 3.11, which shows that the reversible approach leads to a linear increase in the power delivered, respectively, at a constant efficiency, while in the irreversible approach the power curve is a parabola, its maximum corresponding to the optimal speed. The successive consideration of the different irreversibilities (due to the finite speed w, and friction and throttling processes at the intake and exhaust valves) leads to a drastically reduced maximum power output, as well as the corresponding optimal speed (from 150 m/s associated with 145 kW to 20 m/s at approximately 20 kW). As for the efficiency, we observe the same strongly decreasing evolution with the multiplication of the irreversibilities considered.

3.4.4. *Optimization of diesel cycle efficiency at finite speed*

A study similar to that of the previous section was carried out on the irreversible diesel cycle, with isobaric (Petrescu et al. 2009; Petrescu and Costea 2011) and adiabatic combustion, in accordance with the conditions offered by the ceramic engine, a high-efficiency energy solution currently under intensive research.

Figure 3.12 also shows that for the reversible adiabatic cycle the power curve is linear, while for the irreversible cycle the shape is parabolic as a function of the piston speed.

Another important aspect revealed by the curves in Figure 3.12 is that of the reduction in efficiency corresponding to maximum power, from 65% (in the case of the reversible Diesel cycle) to 35%, when all the causes of irreversibility are taken into account. Also, the real values for the piston speed are reduced considerably by the irreversibilities.

Similar results were obtained in the case of the semi–diesel cycle (combustion at constant volume first, then at constant pressure), taking into account the same initial pressure, the same compression ratio and the same heat of combustion for the reversible and irreversible cycles (Petrescu et al. 2009).

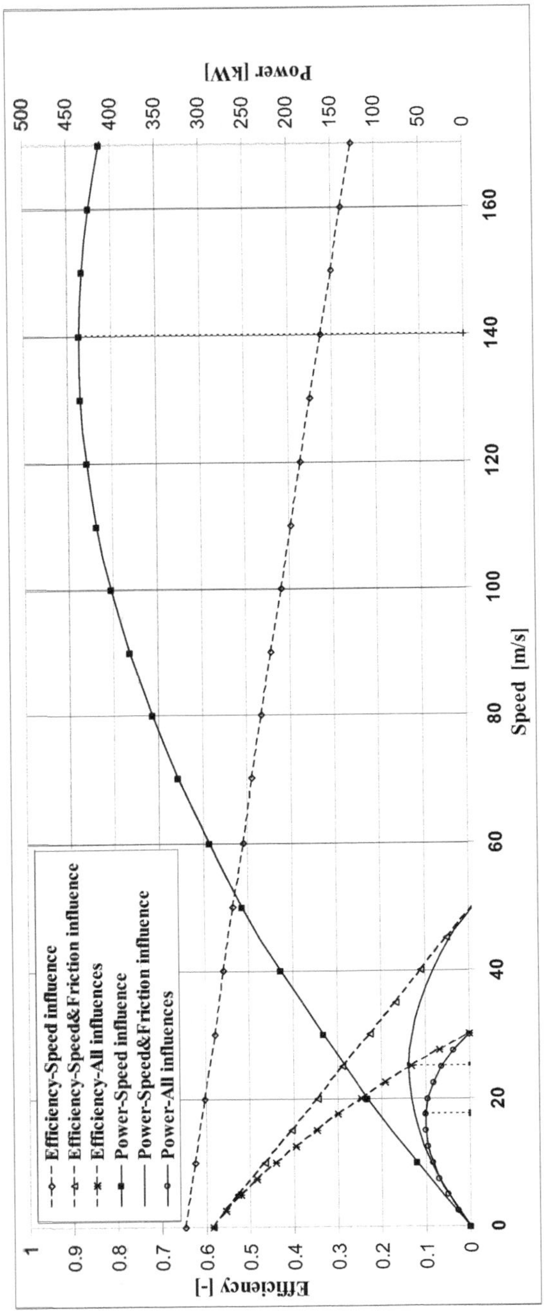

Figure 3.12. *Efficiency and power delivered by the Diesel engine cycle as a function of piston speed (Petrescu et al. 2009; Petrescu and Costea 2011)*

For example, the expression of the efficiency of this cycle is given below, highlighting the contribution of the three causes of irreversibility, finite speed of the processes, throttling in the valves and friction:

$$\eta_{ir,w,lam,f}=$$

$$= 1 - \frac{1}{\varepsilon^{\gamma-1}} \frac{\frac{\alpha_{ir}{}^{k} \cdot \lambda \cdot T_1}{T_1}\left[1 - \frac{a \cdot w}{\sqrt{3R \cdot \alpha_{ir} \cdot \lambda \cdot \varepsilon^{\gamma-1} \cdot T_1}}\left(1 - \left(\frac{\varepsilon}{\alpha_{ir}}\right)^{\frac{\gamma-1}{2}}\right)\right]^2 - 1}{k \cdot \lambda(\alpha - 1) + \lambda - 1}$$

$$- \frac{(\Delta P_f + \Delta P_{thr})\Delta V}{mc_v[(T_3 - T_2) + \gamma(T_4 - T_3)]} \qquad [3.57]$$

By focusing on the most significant work losses, this type of analysis gives the designer the tools to improve and optimize engine efficiency. It paves the way for sensitivity studies revealing the influence of different parameters (temperature, speed, compression ratio, dimensions, etc.) on engine efficiency and the optimization of the engine cycle in relation to these parameters. The internal source of entropy can also be determined and correlated with cycle performance.

3.5. New developments and extension of TFS

There are numerous new developments of TFS using the direct method in various fields, in particular:

– optimization of combined Otto/Stirling cycles (Mc Govern et al. 2010);

– cycles of the steam cold engine (Petrescu et al. 2012);

– comparison between fuel cells and heat engines – a similar approach in the framework of TFS (Petrescu et al. 2013a, 2013b);

– quasi-isothermal liquid piston compressors and expanders (Török et al. 2012);

– thermodynamic tripoles for the study and optimization of irreversible engines (Petrescu et al. 2015a);

– unitary approach to processes in gases, liquids and solids (Petrescu et al. 2015a);

– solar micro-power plant with Fresnel mirrors, Stirling engine, heat storage in oil and electrochemical storage in hydrogen (Petrescu et al. 2010a, 2015a);

– approach to the solar Brayton-Joule cycle and the solar Rankine cycle using TFS (Costea et al. 2014; Martin et al. 2014; Petrescu et al. 2015a).

Encouraged by the achievements mentioned above, we have been looking for new extensions of TFS and the direct method to other fields, different from traditional heat engines.

As many biological systems are in fact electrochemical engines or devices, we tried to use TFS in the study of the most important biological system of human beings (and mammals), the cardio-pulmonary system. The starting point is based on the similarities of this system with certain thermomechanical engines in operation.

Thus, the heart is assimilated to a liquid pump with valves and curved piston for the blood, with the movement of the piston at finite speed throughout life and at variable frequency. The lungs can be thought of as two curved piston air compressors operating in parallel, with the movement of the piston at finite speed throughout life and variable frequency. Then, we can model the system as being composed of "thermo-biological engines" in continuous interaction, designed and built by nature.

The cardio-pulmonary system functions effectively only if a certain phase order of oscillation is established in all stationary states, which assumes:

– a *self-organization of the interaction* between the two pistons of the cardio-pulmonary system;

– this comes from the *entanglement of the two oscillators*, observed in many natural systems, in synergy (Hacken 1983).

By studying the interaction between the heart and the lungs through thousands of measurements of heart rate, lung rate and blood oxygen levels for many people of different ages, we have discovered a surprisingly simple but powerful correlation between the two frequencies in any stationary state, via a *quantum number N*, which has integer value for healthy people, but not integer value for sick people (Petrescu et al. 2014).

The reader will be able to find more information on the development of this new branch of irreversible processes, called *quantum biological thermodynamics with finite speed* in our recent studies (Petrescu et al. 2015a, 2018a, 2018b, 2019a, 2019b; Borcilă et al. 2018, 2019a, 2019b).

Another extension of our research in the field of irreversible thermodynamics has been the attempt to unify thermodynamics in finite physical dimensions (Feidt 2013, 2017, 2018; Blaise et al. 2018) and TFS (Petrescu et al. 2015a, 2015b), as well as that with finite time thermodynamics (Feidt 2009; Petrescu et al. 2015b; Feidt and Costea 2018; Costea et al. 2021).

3.6. International recognition of TFS and the direct method

During the last 10 years, the terms thermodynamics at finite speed, the direct method or the finite speed of the piston have been mentioned in the titles of published articles, which have dealt with the optimization of various heat engines: cold machine and irreversible heat pump (Chen et al. 2011), irreversible Carnot engine (Feng et al. 2011), Robinson engine (Carvajal et al. 2013), Stirling engine (Hosseinzade and Sayyaadi 2015; Hosseinzade et al. 2015; Ahmadi et al. 2016) and irreversible diesel cycle (Wu et al. 2021). Other papers have reported results from approaches that have combined Otto/Stirling cycles (Cullen and McGovern 2010; McGovern et al. 2010).

This is proof that several researchers have recognized the importance of the direct method of TFS and its usefulness in the approach, optimization and design of heat engines with irreversible cycles.

3.7. Conclusion

This chapter has given an overview of the origin and development of TFS and the direct method, as well as the applications, in particular in the field of thermomechanical engines.

The analysis method that we have presented here comes from the fundamental bases of thermodynamics, detailed and systematically developed, from a single equation, that of the *first law of thermodynamics for finite speed processes, combined with the second law*.

The advantage of this method is in the fact that it explains both the causes and the generation mechanisms of irreversibility in complex cycles or actual engines, such as Stirling engines (engines, refrigerators, heat pumps, solar engines), as well as in other cycles such as the Otto, Diesel, Brayton and Carnot cycles.

From the mathematical expression of the first law of thermodynamics for finite speed processes, the direct method makes it possible to analytically obtain the performances of irreversible cycles (efficiency/COP and power) by taking into account the internal and external irreversibilities as a function of a single variable, the finite speed of processes.

This new approach validated the Stirling engine efficiency calculation by comparison with experimental data from a variety of engines. Its application in the study of other thermomechanical systems also provides optimal speeds corresponding to the regimes of maximum performance, these results being useful in their optimized design.

The development of quantum biological thermodynamics with finite speed may help in the design of optimized artificial hearts and artificial lungs, in personalized correlation for each age group of people and their specific characteristics.

3.8. References

Ahmadi, M.H., Ahmadi, M.A., Pourfayaz, F., Bidi, M., Hosseinzade, H., Feidt, M. (2016). Optimization of powered Stirling heat engine with finite speed thermodynamics. *Energy Convers. Manag.*, 108, 96–105.

Allen, D.I. and Tomazic, W.A. (1987). Hot piston ring tests. Report, NASA-TM, 100256, Dearborn, MI.

Bejan, A. (1988). Theory of heat transfer-irreversible power plants. *Int. J. Heat Mass Transfer*, 31(6), 1211–1219.

Blaise, M., Feidt, M., Maillet, D. (2018). Influence of the working fluid properties on optimized power of an irreversible finite dimensions Carnot engine. *Energy Convers. Manag.*, 163, 444–456.

Boltzmann, L. (1898). *Vorlesungen uber Gastheorie – Part I, Part II*. J. A. Barth Verlag, Leipzig.

Borcilă, B., Petrescu, S., Costea, M., Bolohan, R., Petrescu, V., Botez, G. (2018). Effect of flue on stationary states and processes in the cardio-pulmonary system described in quantum biological thermodynamics with finite speed. In *8th International Conference on Advanced Concepts in Mechanical Engineering ACME 2018*. IOP Conference Series: Materials Science and Engineering, Iasi.

Borcilă, B., Petrescu, S., Costea, M., Petrescu, V., Bolohan, R. (2019a). Quantum biological thermodynamics with finite speed of the cardio-pulmonary system II – Computation of power and the entropy source of the cardio-pulmonary system. *Rev. Chim.*, 70(6), 1947–1955.

Borcilă, B., Petrescu, S., Costea, M., Petrescu, V., Bolohan, R. (2019b). Quantum biological thermodynamics with finite speed of the cardio-pulmonary system III – Comparison between stationary states and processes with or without quantum jump for a family members. *Rev. Chim.*, 70(7), 2511–2518.

Carvajal, I., Polupan, G., Jarquin, G., Vazquez, J. (2013). Optimization of the thermal efficiency of a Robinson engine applying the Senft-Schmidt-Petrescu model. *Inf. Tecnol.*, 24(3), 85–94.

Chen, L.G., Feng, H.J., Sun, F.R. (2011). Optimal piston speed ratio analyses for irreversible Carnot refrigerator and heat pump using finite time thermodynamics, finite speed thermodynamics and direct method. *J. Energy Inst.*, 84(2), 105–112.

Costea, M., Grosjean, T., Petrescu, S., Feidt, M., Martin, D., Mladin, E.C., Boriaru, N. (2014). Etude d'un cycle de Rankine solaire à l'aide de la Thermodynamique à Vitesse Finie. In *Colloque Francophone sur l'Energie, Environnement, Economie et Thermodynamique COFRET'14*, Paris.

Costea, M., Petrescu, S., Feidt, M., Dobre, C., Borcilă, B. (2021). Optimization modeling of irreversible Carnot engine from the perspective of combining finite speed and finite time analysis. *Entropy*, 23(5), 504.

Cullen, B. and McGovern, J. (2010). Energy system feasibility study of an Otto cycle/Stirling cycle hybrid automotive engine. *Energy*, 35(2), 1017–1023.

Einstein, A. (1955). *The Meaning of Relativity*. Princeton University Press, NJ.

Farell, R.A., Hindes, C., Battista, R., Connelly, M., Cronin, M., Howarth, R., Donahue, A., Slate, E., Stotts, R., Lacy, R. (1988). Automotive Stirling engine development program. Report, NASA-CR-180839, Cleveland, OH.

Feidt, M. (1987). *Thermodynamique et optimisation énergétique des systèmes et procédés*, 1st edition. Lavoisier TEC & DOC, Paris.

Feidt, M. (2009). Optimal thermodynamics – New upperbounds. *Entropy*, 11(4), 529–547.

Feidt, M. (2013). *Thermodynamique optimale en dimensions physiques finies*. Hermes-Lavoisier, Paris.

Feidt, M. (2017). *Finite Physical Dimensions Optimal Thermodynamics 1: Fundamentals*. ISTE Press, London and Elsevier, Oxford.

Feidt, M. (2018). *Finite Physical Dimensions Optimal Thermodynamics 2: Complex Systems*. ISTE Press, London and Elsevier, Oxford.

Feidt, M. and Costea, M. (2018). From finite time to finite physical dimensions thermodynamics: The Carnot engine and Onsager's relations revisited. *J. Non-Equilib. Thermodyn.*, 43(2), 151–161.

Feng, H.J., Chen, L.G., Sun, F.R. (2011). Optimal ratios of the piston speeds for a finite speed irreversible Carnot heat engine cycle. *Int. J. Sustain. Energy*, 56, 321–335.

Florea, T. (1999). Grapho-analytical method for the study of the operating processes irreversibility in Stirling engines. PhD Thesis, University Politehnica of Bucharest.

Fujii, I. (1990). *From Solar Energy to Mechanical Power*. Harwood Academic Publishers, New York.

Geng, S.M. (1987). Calibration and comparison of the NASA Lewis free-pistons Stirling engine model prediction with RE-1000 test data. Report, NASA-TM-89853, Philadelphia, PA.

Haken, H. (1983). *Advanced Synergetics: Instability Hierarchies of Self-Organizing Systems and Devices*. Springer-Verlag, Berlin, Heidelberg.

Heywood, J.B. (1988). *Internal Combustion Engine Fundamentals*. McGraw-Hill Book Company, New York.

Hosseinzade, H. and Sayyaadi, H. (2015). CAFS: The combined adiabatic-finite speed thermal model for simulation and optimization of Stirling engines. *Energy Convers. Manag.*, 91, 32–53.

Hosseinzade, H., Sayyaadi, H., Babaelahi, M. (2015). A new closed-form analytical thermal model for simulating Stirling engines based on polytropic-finite speed thermodynamics. *Energy Convers. Manag.*, 90, 395–408.

Incropera, F.P. and DeWitt, D.P. (1996). *Fundamentals of Heat and Mass Transfer*. Wiley & Sons, New York.

Macke, W. (1963). *Thermodynamik und Statistik*. Geest & Portig Verlag, Leipzig.

Martin, D., Petrescu, S., Costea, M., Stanciu, C., Feidt, M., Grosjean, T., Boriaru, N. (2014). Étude d'un cycle de Brayton-Joule solaire à l'aide de la Thermodynamique à Vitesse Finie. In *Colloque Francophone sur l'Energie, Environnement, Economie et Thermodynamique COFRET'14*, Paris.

McGovern, J., Cullen, B., Feidt, M., Petrescu, S. (2010). Validation of a simulation model for a combined Otto and Stirling cycle power plant. In *ASME 4th International Conference on Energy Sustainability ES2010*, Phoenix, AZ.

Onsager, L. (1931a). Reciprocal relations in irreversible processes I. *Phys. Rev.*, 37, 405.

Onsager, L. (1931b). Reciprocal relations in irreversible processes II. *Phys. Rev.*, 38, 2265.

Organ, J.A. (1992). *Thermodynamics and Gas Dynamics of Stirling Cycle Machine.* Cambridge University Press, Cambridge.

Paul, M. (1937). Uber die Abhangigkeit der Entropievermehrung quasiadiabatiche Arbaitsprozesse von der Temperatur und Dehnungsgeschwindigkeit. *Ann. D. Phys.*, 5(29), S.179.

Petrescu, S. (1969a). Contribution to the study of thermodynamically non-equilibrium interactions and processes in thermal machines. PhD Thesis, I.P.B., Floreşti.

Petrescu, S. (1969b). An expression for work in processes with finite speed based on linear irreversible thermodynamics. *Studii si Cercetari de Energetica si Electrotehnica*, 19(2), 249–254.

Petrescu, S. (1971a). An elementary deduction of Lorentz-Einstein transformations relations. *Rev. Fizica si Chimie* A, 8(11), 424–430.

Petrescu, S. (1971b). Kinetically consideration regarding the pressure on a movable piston. *Studii si Cercetari de Energetica si Electrotehnica*, 21(1), 93–107.

Petrescu, S. (1973). Study of the gas-gas interaction with finite velocity for flow processes. *Studii si Cercetari de Energetica si Electrotehnica*, 23(2), 99–312.

Petrescu, S. (1974). Experimental study of the gas-piston interaction with finite speed in the case of an open system. *Studii si Cercetari de Mecanica Aplicata*, 31(5), 1081–1086.

Petrescu, S. (1991). Lectures on new sources of energy. Helsinki University of Technology, Otaniemi.

Petrescu, S. and Costea, M. (2011). *Development of Thermodynamics with Finite Speed and Direct Method.* AGIR Publisher, Bucharest.

Petrescu, S. and Harman, C. (1994). The connection between the first and second law of thermodynamics for processes with finite speed. A direct method for approaching and optimization of irreversible processes. *J. Heat Transf. Soc. Jpn.*, 33(128), 60–67.

Petrescu, S., Stanescu, G., Iordache, R., Dobrovicescu, A. (1992). The first law of thermodynamics for closed systems, considering the irreversibilities generated by the friction piston-cylinder, the throttling of the working medium and finite speed of the mechanical interaction. In *Inter. Symposium on Efficiency, Costs, Optimization and Simulation of Energy Systems, ECOS'92*, Zaragoza.

Petrescu, S., Costea, M., Feidt, M. (2001). Les cycles des machines à froid et des pompes à chaleur à vitesse finie. *Revue Entropie*, 232, 51–56.

Petrescu, S., Costea, M., Harman, C., Florea, T. (2002a). Application of the direct method to irreversible Stirling cycles with finite speed. *Int. J. Energy Res.*, 26, 589–609.

Petrescu, S., Feidt, M., Harman, C., Costea, M. (2002b). Optimization of the irreversible Carnot cycle engine for maximum efficiency and maximum power through use of finite speed thermodynamic analysis. In *Proc. ECOS'2002 Conference*, Berlin.

Petrescu, S., Cristea, A.F., Boriaru, N., Costea, M., Petre, C. (2008). Optimization of the irreversible Otto cycle using finite speed thermodynamics and the direct method. In *10th WSEAS Int. Conf. on Mathematical and Computational Methods in Science and Engineering (MACMESE'08), Computers and Simulation in Modern Science*, Bucharest.

Petrescu, S., Boriaru, N., Costea, M., Petre C., Stefan, A., Irimia, C. (2009). Optimization of the irreversible diesel cycle using finite speed thermodynamics and the direct method. *Bul. Transylv. Univ. Brasov*, 2(51), 87–94.

Petrescu, S., Petre, C., Costea, M., Malancioiu, O., Boriaru, N., Dobrovicescu, A., Feidt, M. (2010a). A methodology of computation, design and optimization of solar Stirling power plant using hydrogen/oxygen fuel cells. *Energy*, 35(2), 729–739.

Petrescu, S., Dobre, C., Tirca-Dragomirescu, G., Costea, M., Stanciu, C., Feidt, M. (2010b). Performances evaluation for a reversed quasi-Carnot cycle (refrigeration machine) by using the direct method from finite speed thermodynamics. In *International Conference on Mechanical Engineering, Robotics and Aerospace*, IEEE/IACSIT ICMERA, Bucharest.

Petrescu, S., Dobre, C., Stanciu, C., Costea, M., Tirca-Dragomirescu, G., Feidt, M. (2012). The direct method from thermodynamics with finite speed used for performance computation of quasi-Carnot irreversible cycles I. Evaluation of coefficient of performance and power for refrigeration machines with mechanical compression of vapor. *Rev. Chim.*, 63(1), 74–81.

Petrescu, S., Maris, V., Costea, M., Boriaru, N., Stanciu, C., Dura, I. (2013a). Comparison between the fuel cells and heat engines I. A similar approach in the framework of thermodynamics with finite speed. *Rev., Chim.*, 64(7), 739–747.

Petrescu, S., Maris, V., Costea, M., Boriaru, N., Stanciu, C., Dura, I. (2013b). Comparison between the fuel cells and heat engines II. Operation and performances. *Rev., Chim.*, 64(10), 1187–1193.

Petrescu, S., Petrescu, V., Costea, M., Timofan, L., Danes, S., Botez, G. (2014). Discovery of quantum numbers in the cardio-pulmonary interaction studied in thermodynamics with finite speed. In *Rom. Acad. Tech. Sci Conference Sibiu*, AGIR Publisher, Bucharest.

Petrescu, S., Costea, M., Feidt, M., Ganea, I., Boriaru, N. (2015a). *Advanced Thermodynamics of Irreversible Processes with Finite Speed and Finite Dimensions*. AGIR Publisher, Bucharest.

Petrescu, S., Feidt, M., Enache, V., Costea, M., Stanciu, C., Boriaru, N. (2015b). Unification perspective of finite physical dimensions thermodynamics and finite speed thermodynamics. *Int. J. Energy Environ. Eng.*, 6(3), 245–254.

Petrescu, S., Costea, M., Dobre, C., Borcilă, B., Stanciu, C., Dobrovicescu, A., Popescu, G., Boriaru, N., Banches, E. (2016). Unification between thermodynamics with finite speed and thermodynamics in finite time and its consequences on the development of TFS. In *TE-RE-RD Conference 2016, Golden Sands, Bulgaria*, Politehnica Press, Sofia University.

Petrescu, S., Borcilă, B., Costea, M., Bolohan, R., Petrescu, V., Feidt, M., Botez, G. (2017). Circadian cycles of the cardio-pulmonary system studied in quantum biological thermodynamics with finite speed. *Rev. Termotehnica*, XXI(1), 19–25.

Petrescu, S., Borcilă, B., Costea, M., Bolohan, R., Petrescu, V., Feidt, M., Botez, G. (2018a). About the origin of Stationary States and processes with or without quantum jump in the Cardio-pulmonary system, described in quantum biological thermodynamics with finite speed. *Rev. Termotehnica*, XXII(1), 26–31.

Petrescu, S., Costea, M., Borcilă, B., Petrescu, V., Bolohan, R., Danes S., Danes, F., Feidt, M., Botez, G., Stanescu, G. (2018b). What is quantum biological thermodynamics with finite speed of the cardio-pulmonary system: A discovery or an invention? In *Proc. Rom. Acad. – Math. Phys. Tech. Sci. Inf. Sci.*, 19, 249–254.

Petrescu, S., Borcilă, B., Costea, M., Petrescu, V., Bolohan, R. (2019a). Quantum biological thermodynamics with finite speed of the cardio-pulmonary system I: Fundamental concepts, equations and diagrams. *Rev. Chim.*, 70(5), 1561–1568.

Petrescu, S., Costea, M., Petrescu, V., Bolohan, R., Borcilă, B., Feidt, M. (2019b). Generating structural, interactional, and processual order through synergetic entanglement is the essence for understanding and using quantum biological thermodynamics with finite speed of the cardio-pulmonary system. In *the XXIInd National Conference on Thermodynamics, IOP Conference Series: Materials Science and Engineering*, Galați.

Prigogine, I. (1947). *Etude thermodynamique des phénomènes irréversibles*. Desoer, Liège.

Sommerfeld, A. (1967). *Thermodynamics and Statistical Mechanics* 5. Academic Press, New York.

Stanciu, C. (2007). Utilizarea Termodinamicii cu Viteza Finita in Studiul si Optimizarea ciclului Carnot si a Masinilor Stirling (Utilisation de la Thermodynamique à Vitesse Finie dans l'étude et l'optimisation du cycle de Carnot et des machines de Stirling). PhD Thesis, Politehnica University of Bucharest & Université H. Poincaré Nancy 1.

Stine, W.B. and Diver, R.B. (1994). A compendium of solar dish/Stirling technology. Report, Sandia National Laboratories Report, SAND-93-7026, Albuquerque, NM.

Stoicescu, L. and Petrescu, S. (1964a). The first law of thermodynamics for processes with finite speed, in closed systems. *Bul. IPB*, 26(5), 87–108.

Stoicescu, L. and Petrescu, S. (1964b). Thermodynamic processes developing with constant finite speed. *Bul. IPB*, 26(6), 79–119.

Stoicescu, L. and Petrescu, S. (1965a). Thermodynamic processes with variable finite speed. *Bul. IPB*, 27(1), 65–96.

Stoicescu, L. and Petrescu, S. (1965b). Thermodynamic cycles with finite speed. *Bul. IPB*, 27(2), 82–95.

Stoicescu, L. and Petrescu, S. (1965c). The experimental verification of the new expression of the first law for thermodynamic processes with finite speed. *Bul. IPB*, 27(2), 97–106.

Török, A., Petrescu, S., Popescu, G., Feidt, M. (2012). Compresseurs et détenteurs quasi-isothermes. In *Colloque Francophone sur l'Energie – Environnement – Economie et Thermodynamique COFRET 2012*, Sozopol.

Wu, H., Ge, Y., Chen, L., Feng, H. (2021). Power, efficiency, ecological function and ecological coefficient of performance optimizations of irreversible diesel cycle based on finite piston speed. *Energy*, 216, 119235.

4

Finite Physical Dimensions Thermodynamics

Michel FEIDT

LEMTA, Université de Lorraine, Vandoeuvre-les-Nancy, France

4.1. Introduction

This chapter completes the discussion of the emblematic case of the thermomechanical engine (section 4.2). The approach adopted aims to show the progression of knowledge over the past 50 years, but only on the engine case. The case of reverse cycle machines will be developed in the third volume of this series.

Section 4.2 is a recap of equilibrium thermodynamics, with focus on the CHAMBADAL model. This simplified model will illustrate the entire approach and the current results in the literature.

Section 4.3 will be devoted to the approach of finite-time but endo-irreversible, thermodynamics (Curzon and Ahlborn 1975) different to the endoreversible thermodynamics proposed by Andresen (1983), which constitutes an overestimation of the efficiencies of real engines. Nevertheless, the primitive model remains limited to the configuration without coupling of the heat reservoirs with the converter.

Section 4.5 reports the results obtained with finite physical dimensions optimal thermodynamics (but without coupling). But before approaching this class of generalized model, section 4.4 is dedicated to the finite velocity

Advances in Thermodynamics and Circular Thermoeconomics,
coordinated by Michel FEIDT and Antonio VALERO-CAPILLA.

approach particularly studied by the Romanian school of thermodynamics of the Polytechnic University of Bucharest.

Section 4.6 completes finite physical dimensions optimal thermodynamics, discussing coupling between the reservoirs and the converter. This additional constraint was recently introduced by Michel Feidt during conferences and is the subject of current publications.

The chapter will close with conclusions and perspectives.

4.2. The Carnot engine according to equilibrium thermodynamics

4.2.1. *Equilibrium thermodynamics and heat losses*

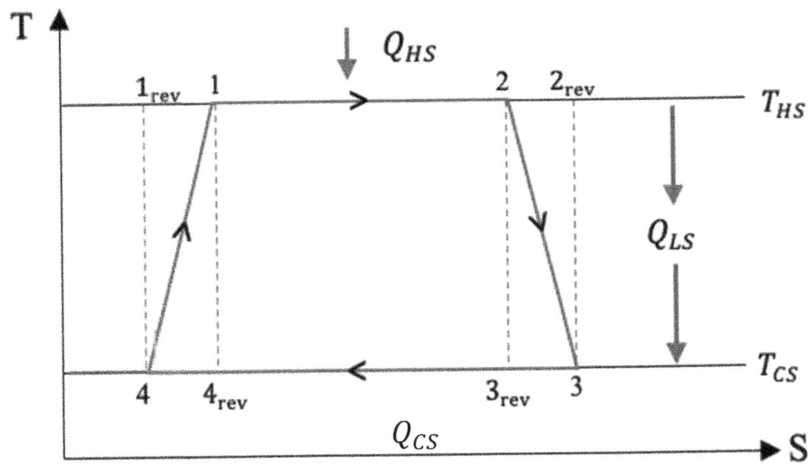

Figure 4.1. *Carnot cycle with irreversibility of the converter. For a color version of this figure, see www.iste.co.uk/feidt/advances.zip*

In studies according to equilibrium thermodynamics, the coupling of the reservoirs with the converter is irrelevant.

Here, the sources and sinks are thermostats at the respective temperatures T_{HS} and T_{HC}.

The study of the cycle is carried out in energy, with all the following quantities being positive:

– Q_H is the heat energy received by the converter at the hot end;

– Q_C is the heat energy rejected by the converter at the cold end;

– W is the mechanical energy produced in the converter.

Energy conservation then provides the energy balance at the converter:

$$W = Q_H - Q_C \quad [4.1]$$

The non-adiabaticity of the system (source – converter – sink) results in the overall heat losses between the hottest point and the coldest point Q_{LS} such that the energetic cost is given as:

$$Q_{HS} = Q_H + Q_{LS} \quad [4.2]$$

Heat rejection:

$$Q_{CS} = Q_C + Q_{LS} \quad [4.3]$$

For a reversible Carnot cycle:

– if the reference is the hot end (imposed transformation 1–2), the reversible cycle is represented by 1–2–3_{rev}–4_{rev}–1;

– if the reference is the cold end (imposed transformation 3–4), the reversible cycle is represented by 1_{rev}–2_{rev}–3–4.

In both cases, the entropy balance is written as:

$$\frac{Q_H}{T_{HS}} = \frac{Q_C}{T_{CS}} \quad [4.4]$$

By recalling that the efficiency of the engine in the sense of the first law is the efficiency η_I, for the converter, it becomes:

$$\eta_{IC} = \frac{W}{Q_H} = 1 - \frac{Q_C}{Q_H} = 1 - \frac{T_{CS}}{T_{HS}} \quad [4.5]$$

the well-known Carnot efficiency, but at zero power since it is subject to four quasistatic transformations, hence a cycle of infinite duration.

If heat losses are taken into account, by combining equations [4.1]–[4.4], it becomes:

$$\eta_{IL} = 1 - \frac{Q_{CS}}{Q_{HS}} = 1 - \frac{T_{CS}}{T_{HS}} \frac{1+\frac{Q_{LS}}{Q_C}}{1+\frac{Q_{LS}}{Q_H}} \qquad [4.6]$$

It should be noted that the presence of heat losses affects the efficiency, and introduces a production of entropy for the system (source–converter–sink) ΔS_{IS}:

$$\Delta S_{IS} = Q_{LS}\left(\frac{1}{T_{CS}} - \frac{1}{T_{HS}}\right) \qquad [4.7]$$

4.2.2. *Exo-reversible Carnot engine (endo-irreversible)*

The corresponding model uses the previous equations [4.1]–[4.3], but the relation [4.4] becomes:

$$\frac{Q_H}{T_{HS}} + \Delta S_I = \frac{Q_C}{T_{CS}} \qquad [4.8]$$

where ΔS_I is the total entropy production over the cycle at the converter.

We deduce the mechanical energy by combining [4.1] and [4.8]:

$$W = Q_H\left(1 - \frac{T_{CS}}{T_{HS}}\right) - T_{CS}\Delta S_I \qquad [4.9]$$

Then the corresponding efficiency of the converter is given as:

$$\eta_I = \eta_C - \frac{T_{CS}\Delta S_I}{Q_H} \qquad [4.10]$$

Finally, the corresponding efficiency of the system is given as:

$$\eta_{IS} = \eta_C \frac{Q_H}{Q_{HS}} - \frac{T_{CS}\Delta S_I}{Q_{HS}} \qquad [4.11]$$

It can therefore be seen that the efficiency of the system is reduced, due to heat losses and internal irreversibilities in the converter. The maximum energy efficiency implies a minimization of heat losses (adiabaticity) and the minimization of entropy production at the converter in accordance with the Gouy–Stodola theorem.

4.2.3. *Power of the exo-reversible Carnot engine*

By taking the expression [4.9] for the mechanical energy and dividing it by τ, the cycle period, the average mechanical power of the exor-reversible engine is obtained:

$$\bar{\dot{W}} = \eta_C \frac{Q_H}{\tau} - \frac{T_{CS}\Delta S_I}{\tau} \qquad [4.12]$$

Note that $\eta_C Q_H$ represents the reversible mechanical work associated with the finite thermal energy Q_H:

$$\bar{\dot{W}} = \frac{W_{rev}}{\tau}\left(1 - \frac{T_{CS}\Delta S_I}{W_{rev}}\right) \qquad [4.13]$$

As a result, at imposed $\bar{\dot{W}}_{rev}$, the optimum power remains associated with the reversibility of the engine.

In contrast, in *finite energy* (W_{rev} imposed), the maximum cycle average power occurs for a given period τ^*, assuming $\Delta S_I = \frac{C_I}{\tau}$ (*C_I action of entropy production*) such that:

$$\tau^* = \frac{2T_{CS}C_I}{W_{rev}} \qquad [4.14]$$

and:

$$MAX\ \bar{\dot{W}} = \frac{1}{4}\frac{W_{rev}^2}{T_{CS}C_I} \qquad [4.15]$$

which corresponds to:

$$\eta_I\left(MAX\ \bar{\dot{W}}\right) = \frac{\eta_C}{2} \qquad [4.16]$$

At this stage of the model, we can introduce an *efficiency in the sense of the second law* η_{II}, by relating engine efficiency to Carnot efficiency:

$$\eta_{II} = \frac{\eta_I}{\eta_C} \qquad [4.17]$$

This efficiency represents a quality factor of the installation (Feidt 1987).

The particularizations of the converter, then of the system, lead to:

$$\eta_{II} = 1 - \frac{T_{CS}\Delta S_I}{\eta_C Q_H} = 1 - \frac{T_{CS}\Delta S_I}{W} \quad [4.18]$$

$$\eta_{II\,S} = \frac{Q_H}{Q_{HS}} - \frac{T_{CS}\Delta S_I}{\eta_C Q_{HS}} \quad [4.19]$$

4.3. The Chambadal engine model (1957)

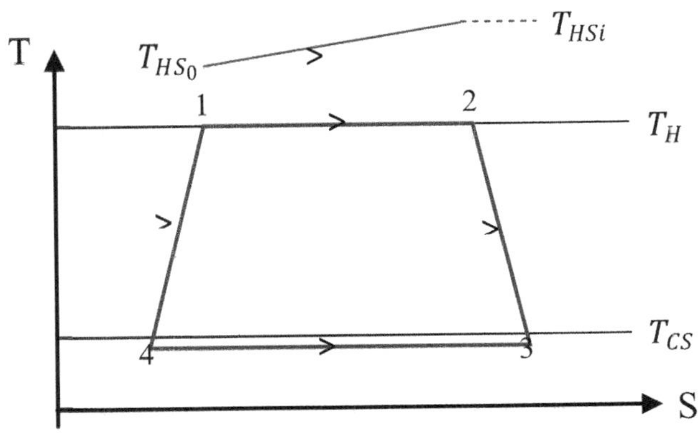

Figure 4.2. *Chambadal engine with irreversibility at the converter. For a color version of this figure, see www.iste.co.uk/feidt/advances.zip*

The model is intermediate between the previous model and the Curzon–Ahlborn model (Curzon and Ahlborn 1975). The original model assumes a stationary steady-state regime of the system, and thus a reasoning in heat flux as well as in power for the engine; we then return to the development carried out in section 4.2.3, but with the assumption of a perfect thermal contact with the sink, while the source becomes a finite source, with heat flow $\dot{c}_H = \dot{m}_H c_{pH}$, where $\dot{m}_H$ is the mass flow rate of the source fluid, c_{pH} is the specific mass heat of the source fluid, and the entering temperature is T_{HSi}. The model, as we have shown, has a much older history, and is still in an endoreversible configuration (Vaudrey et al. 2014; Feidt 2017).

Here, we extend the model to the endo-irreversible case, so that the system model becomes irreversible except at the heat sink, but remains adiabatic because there are no heat losses.

In addition to the internal irreversibility of the converter ΔS_I, the heat transfer at the hot end appears to be irreversible. However, we will consider, for the sake of simplification, a system without heat losses, as mentioned above.

Using the method of efficiency and numbers of transfer units (NTU), the expression of the heat flux transferred to the hot end (Feidt 2013) becomes:

$$\dot{Q}_H = \varepsilon_H \dot{c}_H (T_{HSi} - T_H) \qquad [4.20]$$

From this, we deduce for a cycle of duration τ, the expression associated with thermal energy Q_H:

$$Q_H = \varepsilon_H \dot{c}_H \tau (T_{HSi} - T_H) \qquad [4.21]$$

It is then convenient to introduce the finite physical dimension at the hot end, in the form of a generalized heat conductance G_H, such that:

$$Q_H = G_H (T_{HSi} - T_H) \qquad [4.22]$$

Note that for a Chambadal engine, there is a second expression of Q_H:

$$Q_H = T_H \Delta S_H \qquad [4.23]$$

This expression will be used for optimization in the presence of source-converter coupling.

4.3.1. *Optimization of the modified Chambadal engine without source-converter heat transfer coupling constraint*

In Feidt and Costea (2019a, 2019b), the optimization is carried out by two methods: the entropy ratio method and then the entropy production method by analyzing the influence of the form of the entropy production law.

We will limit the discussion here to a constant parameter entropy production, ΔS_I.

The expression of the mechanical energy [4.9] then becomes:

$$W = G_H(T_{HSi} - T_H)\left(1 - \frac{T_0}{T_H}\right) - T_0 \Delta S_I \quad [4.24]$$

with $T_{CS} = T_0$.

The maximum mechanical energy is then:

$$MAX_1(W) = G_H\left(\sqrt{T_{HSi}} - \sqrt{T_0}\right)^2 - T_0 \Delta S_I \quad [4.25]$$

The associated efficiency is $\eta_I(MAX_1(W))$:

$$\eta_I(MAX_1(W)) = 1 - \sqrt{\frac{T_0}{T_{HSi}}} - \frac{\Delta S_I}{G_H}\frac{T_0}{T_{HSi} - \sqrt{T_{HSi}T_0}} \quad [4.26]$$

We then see in the previous relation the efficiency of the endoreversible engine with a finite energy source:

$$\eta_{endo}(MAX_1(W)) = 1 - \sqrt{\frac{T_0}{T_{HSi}}} \quad [4.27]$$

The previous expression is a generalization of the efficiency established for an infinite source and sink (thermostats) by Curzon and Ahlborn (1975). This efficiency is an upper limit.

In the presence of internal irreversibility, the maximum efficiency will be associated with the minimum of the ratio $\frac{\Delta S_I}{G_H}$, at given temperatures (T_{HSi}, T_0).

If there is a finite energy source G_H, assuming the production of entropy is again inversely proportional to the cycle duration τ, it is easy to find that the maximum average power occurs for a period τ^* (see Appendix):

$$\tau^* = \frac{2T_0C_I}{W_{endo}} \quad [4.28]$$

with $W_{endo} = G_H\left(\sqrt{T_{HSi}} - \sqrt{T_0}\right)^2$

$MAX_1(W)$ equals $\frac{W_{endo}}{2}$, and therefore:

$$MAX_1\left(\bar{W}\right) = \frac{W_{endo}^2}{4T_0C_I} \quad [4.28']$$

and:

$$\eta_I(MAX_1(W)) = \eta_I\left(MAX_1\left(\bar{W}\right)\right) = \frac{1}{2}(1 - \sqrt{\frac{T_0}{T_{HSi}}}) \quad [4.28'']$$

These results are new results from the literature.

4.3.2. *Optimization of the modified Chambadal engine, with the transfer entropy constraint*

The comparison of relations [4.22] and [4.23] shows that the coupling between the source and the converter imposes an interdependence of G_H characterizing the source (ΔS_H), which is an existence condition of the cycle, and T_H, the temperature of the cycled fluid. This results in:

$$T_H = \frac{G_H}{G_H+\Delta S_H}T_{HSi} \quad [4.29]$$

For the engine, it is logical to take as reference the heat transfer entropy at the input of the converter (ΔS_H), before the conversion of heat into work.

The combination of [4.29] with [4.24] gives:

$$W = \frac{\Delta S_H}{G_H+\Delta S_H}[G_H(T_{HSi} - T_0) - T_0\Delta S_H] - T_0\Delta S_I \quad [4.30]$$

The variable ΔS_H is optimized by derivation, resulting in a quadratic equation whose physical root is:

$$\Delta S_H^* = G_H(\sqrt{\frac{T_{HSi}}{T_0}} - 1) \quad [4.31]$$

This corresponds to:

$$T_H^* = \sqrt{T_{HSi}T_0} \quad [4.32]$$

This result restores the same energy maximum [4.25] and the associated efficiency [4.26].

The successive power optimization will therefore be identical (relations [4.28], [4.28'] and [4.28"]). The results are therefore generic for the Chambadal model. Other relevant cases include (Feidt and Costea 2019a, 2019b; Feidt and Feidt 2021).

4.3.3. *Section conclusion*

If the models of thermomechanical engines started with the thermostatic model, they quickly evolved toward models in stationary steady-state regime (Vaudrey et al. 2014; Feidt 2017).

In these models, we find transfers by latent heat (Carnot model), as well as by sensible heat (Chambadal model) (Chambadal 1957).

If thermostatics provides access to an upper limit for efficiency in the sense of the first law, this limit corresponds to an engine delivering zero power.

The Chambadal model, working in a stationary steady-state regime with a finite physical source, both in energy (G_H) and flux ($\dot{c}_H$), allows maximum mechanical energy to be obtained W_{endo} (with or without coupling of source and converter). In the presence of internal irreversibility characterized by the action of entropy production C_I, note the existence of an optimal cycle period leading to maximum average power. A remarkable result is that the efficiency at this maximum power is half the efficiency obtained in the endoreversible case (relation [4.28"]). This completes the set of results reported in Feidt (2013).

4.4. The Curzon–Ahlborn model (1975)

This model, unlike the Chambadal model, preserves the formal symmetry between the hot end and the cold end. It is an adiabatic model (without heat loss) and endoreversible (no internal irreversibility in the converter on the cycle, see Figure 4.3).

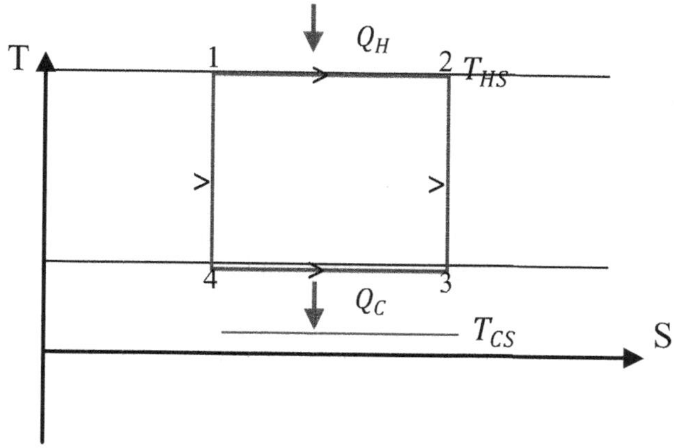

Figure 4.3. *Endoreversible configuration of Curzon–Ahlborn engine (according to original article). For a color version of this figure, see www.iste.co.uk/feidt/advances.zip*

The reasoning of Curzon–Ahlborn is based on the four transformations of the cycle, which are reversible:

– (1 – 2) reversible isothermal expansion;

– (2 – 3) reversible adiabatic expansion;

– (3 – 4) reversible isothermal compression;

– (4 – 1) reversible adiabatic compression.

The only irreversibilities of the system (hot source–thermomechanical converter–cold source) are the irreversibilities of heat transfers.

As we will see, the authors then use heat transfer transformation conductances K'_H, K'_C.

The heat transfer fluxes therefore expressed the form:

$$\dot{Q}_i = K'_i(T_{Si} - T_i) \quad [4.33]$$

Linearity assumption, with $K'_i = const.$

Whereby:

$$\dot{Q}_H = K'_H(T_{HS} - T_H) \quad [4.34]$$

$$\dot{Q}_C = K'_C(T_C - T_{CS}) \quad [4.35]$$

The finite time dimension (duration) then appears in the expressions of the quantities of heat exchanged at the hot end Q_H (duration t_H) and cold end (duration t_C):

$$Q_H = K'_H t_H(T_{HS} - T_H) \quad [4.36]$$

$$Q_C = K'_C t_C(T_C - T_{CS}) \quad [4.37]$$

4.4.1. *Optimization of mechanical energy*

The mechanical energy of the engine is given as a function of four variables [4.1]:

$$W = Q_H - Q_C = f(T_H, T_C, t_H, t_C) \quad [4.38]$$

$T_{HS}, T_{CS}, K'_H, K'_C$ are the parameters.

Taking into account the entropy constraint [4.4], and using variational calculation (Feidt 1996), we find the optimal state point of the cycle described by the temperatures T^*_H, T^*_C:

$$T^*_H = \sqrt{T_{HS}}\left(\frac{K'_H t_H\sqrt{T_{HS}}+K'_C t_C\sqrt{T_{CS}}}{K'_H t_H+K'_C t_C}\right) \quad [4.39]$$

$$T^*_C = \sqrt{T_{CS}}\left(\frac{K'_H t_H\sqrt{T_{HS}}+K'_C t_C\sqrt{T_{CS}}}{K'_H t_H+K'_C t_C}\right) \quad [4.40]$$

We deduce the optimal cycle work $MAX_1(W)$:

$$MAX_1(W) = \frac{K'_H t_H K'_C t_C}{K'_H t_H+K'_C t_C}\left(\sqrt{T_{HS}} - \sqrt{T_{CS}}\right)^2 \quad [4.41]$$

If we determine $Q_H(MAX_1(W))$, the expression of the efficiency in the sense of the first law results in:

$$\eta_I\left(MAX_1(W)\right) = 1 - \sqrt{\frac{T_{CS}}{T_{HS}}} \qquad [4.42]$$

This efficiency corresponds to the efficiency of an endoreversible Carnot engine at maximum energy and the result obtained in [4.27], but here for thermostatic sources and sinks.

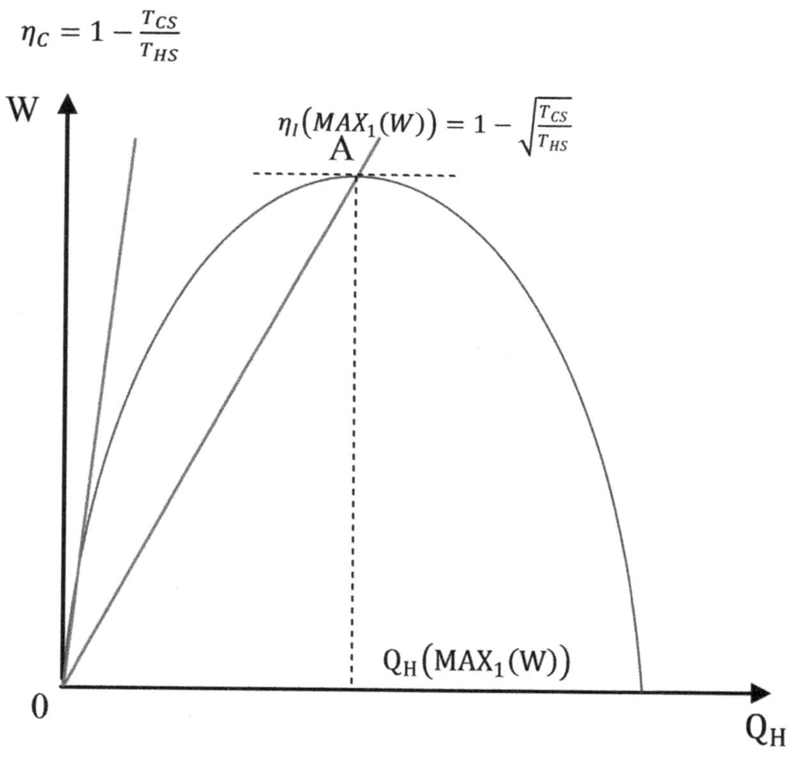

Figure 4.4. *Evolution of work W as a function of Q_H for the endoreversible Carnot engine. For a color version of this figure, see www.iste.co.uk/feidt/advances.zip*

An important finding is the relationship of the various parameters to each other: we note that $K_i' t_i$ is a thermal energy transfer conductance G_i, but written on the basis of isothermal transformations. If we reason on cycle averages, G_i becomes $G_i = K_i.\tau$, with K_i as the classical conductance of thermal energy transfer such that $K_i = k_i A_i$, with k_i as the overall heat transfer coefficient and A_i as the heat transfer area considered. This approach is preferable compared to that of Curzon–Ahlborn, because it is closer to the studied physical reality; it accounts for operation in a steady-state regime as in the Chambadal model.

Finally, if we seek to optimize the cycle efficiency, we find the Carnot efficiency limit (see Figure 4.4).

4.4.2. Optimization of mechanical power

The total cycle period τ is the sum of the durations of each transformation:

$$\tau = t_H + t_{Ex} + t_C + t_{C_0} \quad [4.43]$$

where t_{Ex} is the isentropic expansion time, t_{C_0} is the isentropic compression time.

The Curzon–Ahlborn model assumes that the mechanical transfers are much faster than the heat transfers, but that these durations are nevertheless proportional:

$$\tau = \alpha(t_H + t_C) \quad [4.44]$$

with $\alpha = 1 + \varepsilon$, ε being small.

The expression for the average power over the cycle is deduced from this:

$$\bar{W} = \frac{K'_H t_H (T_{HS} - T_H) - K'_C t_C (T_C - T_{CS})}{\alpha(t_H + t_C)} \quad [4.45]$$

$\bar{W}$ is therefore a function of four variables (T_H, T_C, t_H, t_C) linked by the entropy constraint [4.4].

The variational calculation then makes it possible to obtain the maximum power $MAX(\bar{W})$ and the corresponding state point as follows:

$$T_H^* = \sqrt{T_{HS}}\left(\frac{\sqrt{K'_H T_{HS}} + \sqrt{K'_C T_{CS}}}{\sqrt{K'_H} + \sqrt{K'_C}}\right) \quad [4.46]$$

$$T_C^* = \sqrt{T_{CS}}\left(\frac{\sqrt{K'_H T_{HS}}+\sqrt{K'_C T_{CS}}}{\sqrt{K'_H}+\sqrt{K'_C}}\right) \quad [4.47]$$

$$MAX\left(\bar{\dot{W}}\right) = \frac{K'_H K'_C}{\alpha}\frac{\left(\sqrt{T_{HS}}-\sqrt{T_{CS}}\right)^2}{\left(\sqrt{K'_H}+\sqrt{K'_C}\right)^2} \quad [4.48]$$

A new optimization then becomes possible, noting that the thermal conductances of the isothermal transformations must be finite. We can therefore seek the optimal distribution of these, limited by the following finiteness constraint:

$$K'_H + K'_C = K'_T \quad [4.49]$$

This will preferably be done on K_H and K_C in a planned publication for the scientific journal entropy.

4.5. Finite speed thermodynamics

4.5.1. *Introduction*

For a fluid like a liquid or better yet a gas, the movements of molecules can transmit impulses via successive collisions. At the macroscopic level, the molecular pulses generate pressure waves causing the wall of the volume to vibrate (cylinder in an engine).

This corresponds to sound waves that lead to an out of equilibrium state. The speed of sound waves is about 340 m/s (lower than the speed of fluid molecules). This value is valid for sound waves in air (or diatomic gas) at a temperature of around 15°C.

It is this non-equilibrium state that is the focus of the study by Petrescu et al. (2015) at the Polytechnic University of Bucharest (Romania) on finite speed thermodynamics.

4.5.2. *Expression of the first law of thermodynamics for finite speed processes*

Here, note the differential expression modified to satisfy the fundamental law of finite speed thermodynamics, namely:

$$dU = \delta Q_{irr} + \delta W_{irr} \quad [4.50]$$

with:

$$\delta W_{irr} = -P_{m,i}\left[1 \pm \frac{aw}{c}\right] dV \quad [4.51]$$

where w is the mean piston speed, c is the average molecular speed ($c = \sqrt{3RT}$):

$$a = \sqrt{3\frac{c_p}{c_v}}$$

The sign + corresponds to the compression of the fluid, and the sign – corresponds to the expansion of the fluid.

4.5.3. *Extension of the method*

This method has been successfully applied to the Otto, Carnot and Stirling engines and to reverse cycle engines. The reader will be able to find more substantial details and developments in Chapter 3 of this book and in Petrescu et al. (2015).

The method used to analyze finite speed irreversible processes, called the direct method, assumes the integration of the new mathematical expression of the first law on each type of irreversible process, resulting in modified equations of the processes, which lead to an analytical study of the efficiency of thermal engines.

Note in particular that the authors introduce the internal irreversibilities by the loss separation method. The approach to pressure losses particularly considers those by friction, on the one hand, and by throttling, on the other hand, in closed systems. These losses are added in the dimensionless form in [4.51] to the term *aw/c*, responsible for the pressure losses due to the non-uniformity of pressure in the cylinder.

Analytical expressions for friction and throttling pressure losses are derived from an empirical relationship (Heywood 1983) based on experimental data from internal combustion engines:

$$\Delta P_{total} = 0.97 + 0.15\frac{N_r}{1{,}000} + 0.05\left(\frac{N_r}{1{,}000}\right)^2 \text{ [bar]} \qquad [4.52]$$

which has been transformed as a function of the piston speed and separated by irreversibility as follows:

$$\Delta P_f = 0.97 + 0.45\frac{w}{100 \cdot S}; \quad \Delta P_{lam} = 0.45\left(\frac{w}{100 \cdot S}\right)^2 \text{ [bar]} \qquad [4.53]$$

These expressions will change again for the Stirling engine, considered as a complex closed system, because of the presence of a regenerative heat exchanger. More details on these pressure losses are given in Chapter 3 of this book.

4.6. Finite physical dimensions optimal thermodynamics: case without coupling of sources and sinks with the converter

Again, we consider the Curzon–Ahlborn model as reported in section 4.4 in the adiabatic and endoreversible case. The energy optimization revealed a generalized thermal energy transfer conductance G_i:

$$G_i = K_i' \tau_i = K_i \tau \qquad [4.54]$$

such that the maximum work of the endoreversible engine is written as:

$$MAX_1(W_{endo}) = \frac{G_H G_C}{G_H + G_C}\left(\sqrt{T_{HS}} - \sqrt{T_{CS}}\right)^2 \qquad [4.55]$$

If we introduce the finiteness constraint on G_i:

$$G_H + G_C = G_T \qquad [4.56]$$

It is easy to deduce $MAX_2(W_{endo})$, which corresponds to the equipartition of the generalized conductances:

$$G_H^* = G_C^* = \frac{G_T}{2} \qquad [4.57]$$

and:

$$MAX_2(W_{endo}) = \frac{G_T}{4}\left(\sqrt{T_{HS}} - \sqrt{T_{CS}}\right)^2 \qquad [4.58]$$

We also used this calculation in the irreversible case for a finite source and sink with respective inlet temperatures T_{HSi}, T_{CSi}. By stating $G_i = \varepsilon_i \dot{C}_i \tau$, [4.24] becomes:

$$W = G_H(T_{HSi} - T_H) - G_C(T_C - T_{CSi}) \qquad [4.59]$$

With the entropy constraint:

$$\frac{G_H(T_{HSi}-T_H)}{T_H} + \Delta S_I = \frac{G_C(T_C-T_{CSi})}{T_C} \qquad [4.60]$$

Assuming $T_{HSi}, T_{CSi}, G_H, G_C, \Delta S_i$ are parameters, and T_H, T_C are variables, we find the first engine work optimum for the state vector:

$$T_H^* = \sqrt{T_{HSi}}\frac{(G_H\sqrt{T_{HSi}}+G_C\sqrt{T_{CSi}})}{G_T-\Delta S_I} \qquad [4.61]$$

$$T_C^* = \sqrt{T_{CSi}}\frac{(G_H\sqrt{T_{HSi}}+G_C\sqrt{T_{CSi}})}{G_T-\Delta S_I} \qquad [4.62]$$

and:

$$MAX_1(W) = \frac{G_H G_C}{G_T-\Delta S_I}\left(\sqrt{T_{HS}} - \sqrt{T_{CS}}\right)^2 - \frac{\Delta S_I}{G_T-\Delta S_I}(G_H T_{HSi} + G_C T_{CSi}) \qquad [4.63]$$

We also see, in [4.63], that $MAX_1(W)$ is a decreasing function of ΔS_I, according to the Gouy–Stodola theorem.

The maximum efficiency of mechanical work is expressed in the form:

$$\eta_I(MAX_1(W)) = 1 - \sqrt{\frac{T_{CS}}{T_{HS}}\frac{1+\frac{\Delta S_I}{G_H}\frac{\sqrt{T_{CS}}}{\sqrt{T_{HS}}-\sqrt{T_{CS}}}}{1-\frac{\Delta S_I}{G_C}\frac{\sqrt{T_{HS}}}{\sqrt{T_{HS}}-\sqrt{T_{CS}}}}} \qquad [4.64]$$

There is therefore a factor F accounting for the decrease in the maximum efficiency of mechanical work such that:

$$\eta_I\big(MAX_1(W)\big) = 1 - \sqrt{\frac{T_{CS}}{T_{HS}}}F \qquad [4.65]$$

The endoreversible efficiency is indeed an upper bound to the efficiency with source and sink determined by G_H, G_C. But $\eta_I\big(MAX(W)\big)$ is an increasing function of G_H and decreasing function of G_C. There is therefore an optimal distribution of generalized finite-dimensional conductances to obtain maximum efficiency or maximum mechanical energy.

We thus find in the latter case the following optimal distribution:

$$G_H^* = \frac{1}{2}\left[G_T - \Delta S_I \frac{\sqrt{T_{HS}}+\sqrt{T_{CS}}}{\sqrt{T_{HS}}-\sqrt{T_{CS}}}\right] \qquad [4.66]$$

$$G_C^* = \frac{1}{2}\left[G_T + \Delta S_I \frac{\sqrt{T_{HS}}+\sqrt{T_{CS}}}{\sqrt{T_{HS}}-\sqrt{T_{CS}}}\right] \qquad [4.67]$$

and:

$$MAX_2(W) = \frac{1}{4(G_T-\Delta S_I)}\left[G_T^2\left(\sqrt{T_{HS}} - \sqrt{T_{CS}}\right)^2 - 2G_T\Delta S_I(T_{HS} + T_{CS}) + 2\Delta S_I^2\left(\sqrt{T_{HS}} + \sqrt{T_{CS}}\right)^2\right] \qquad [4.68]$$

It is confirmed that for equations [4.66] and [4.67] the equipartition only corresponds to the endoreversible case ($\Delta S_I = 0$). In all other cases, given the positivity of the function ΔS_I, G_C^* is greater than G_H^*.

Finally, power optimization, given the choice of diagram in $G_i = K_i\tau$, shows that $\bar{\dot{W}}$, the average power at $MAX_1(W)$, has as its maximum value that associated with the equipartition of K_i:

$$MAX\left(\bar{\dot{W}}\right) = \frac{K_T}{4}\left(\sqrt{T_{HS}} - \sqrt{T_{CS}}\right)^2 \qquad [4.69]$$

4.7. Finite physical dimensions optimal thermodynamics: case with source and sink coupling with the converter

This model is inspired by that of the adiabatic Curzon–Ahlborn model and extended in section 4.6, also introducing an existence constraint of the irreversible cycle by a reference heat transfer entropy, in addition to the total entropy production in the cycle ΔS_I.

The objective is still to obtain maximum work of the engine, but with the coupling constraints at the hot end and at the cold end, namely:

$$Q_H = G_H(T_{HS} - T_H) = T_H \Delta S_H = T_{HS} \Delta S_{HS} \quad [4.70]$$

$$Q_C = G_C(T_C - T_{CS}) = T_C \Delta S_C = T_{CS} \Delta S_{CS} \quad [4.71]$$

with $\Delta S_{HS} \leq \Delta S_H \leq \Delta S_C \leq \Delta S_{CS}$ for heat transfer entropies, and corresponding heat intensities: $T_{HS} \geq T_H \geq T_C \geq T_{CS}$.

The entropy production during the heat transfer between the source and the converter is equal to ΔS_{IH}:

$$\Delta S_{IH} = \Delta S_H - \Delta S_{HS} = Q_H\left(\frac{1}{T_H} - \frac{1}{T_{HS}}\right) \quad [4.72]$$

Similarly, during the heat transfer between the converter and the cold end, the production of entropy ΔS_{IC} is:

$$\Delta S_{IC} = \Delta S_{CS} - \Delta S_C = Q_C\left(\frac{1}{T_{CS}} - \frac{1}{T_C}\right) \quad [4.73]$$

The entropy balance of the engine cycle is updated from [4.8] in the form:

$$\frac{Q_H}{T_H} + \Delta S_i = \frac{Q_C}{T_C} \quad [4.74]$$

The choice of reference heat transfer entropy for an engine is between ΔS_{HS} and ΔS_H. We retain the transfer entropy at the engine input, $\Delta S_H = \Delta S$.

In accordance with the linear heat transfer law expressed in [4.70] and [4.71], the related variables $T_H, T_C, \Delta S$ are such that:

$$T_H = T_{HS} \frac{G_H}{G_H + \Delta S} \quad [4.75]$$

$$T_C = T_{CS} \frac{G_C}{G_C - (\Delta S + \Delta S_I)} \quad [4.76]$$

The cycle temperatures are therefore a function of G_H, with G_C as the heat transfer conductances at the hot and cold ends of the system, but also of the main extensive variable of the converter ΔS, and its irreversibilities ΔS_I.

4.7.1. *Irreversible Carnot engine with finite source and sink*

Heat transfer conductances G_H, G_C have known parameters.

The temperature inequality constraints expressed at the beginning of the section then impose the following for the converter variable ΔS:

$$\Delta S < \frac{(T_{HS}-T_{CS})G_C - T_{HS}G_H\Delta S_I}{G_H T_{HS} + G_C T_{CS}} \qquad [4.77]$$

The expression of the work of the mechanical cycle is then written as:

$$W = T_{HS}\frac{G_H\Delta S}{G_H+\Delta S} - T_{CS}\frac{G_C(\Delta S+\Delta S_I)}{G_C+(\Delta S+\Delta S_I)} \qquad [4.78]$$

Keeping the assumption that ΔS_I is a constant parameter, by derivation from W as a function of the variable ΔS, we obtain $MAX_1(W)$ the maximum mechanical energy for the configuration ΔS^* of the converter:

$$\Delta S^* = \frac{G_H G_C\left(\sqrt{T_{HS}}-\sqrt{T_{CS}}\right) - G_H\Delta S_I\sqrt{T_{HS}}}{G_H\sqrt{T_{HS}}+G_C\sqrt{T_{CS}}} \qquad [4.79]$$

with:

$$MAX_1(W) = \frac{1}{G_T-\Delta S_I}\left[G_H G_C\left(\sqrt{T_{HS}}-\sqrt{T_{CS}}\right)^2 - \Delta S_I(G_H T_{HS} + G_C T_{CS})\right] \qquad [4.80]$$

We find the same formalism as at the beginning of the section (relation [4.63]) and therefore the same conclusions, if we assume G_H, G_C are parameters, and G_H, G_C are variables (relations [4.66]–[4.69]).

4.7.2. *Irreversible Carnot engine with finite converter*

In this case, the reference heat transfer entropy ΔS, which describes the existence of the cycle, is a constant parameter. T_H and T_C remain consistent with [4.75] and [4.76].

The aim is to obtain the maximum mechanical work, with the finite dimensional constraint on the variables G_H and G_C. It becomes:

$$G_H^* = \Delta S \frac{G_T\sqrt{T_{HS}}-(\Delta S+\Delta S_I)(\sqrt{T_{HS}}+\sqrt{T_{CS}})}{\Delta S\sqrt{T_{HS}}+(\Delta S+\Delta S_I)\sqrt{T_{CS}}} \quad [4.81]$$

$$G_C^* = (\Delta S + \Delta S_I)\frac{G_T\sqrt{T_{CS}}+\Delta S(\sqrt{T_{HS}}+\sqrt{T_{CS}})}{\Delta S\sqrt{T_{HS}}+(\Delta S+\Delta S_I)\sqrt{T_{CS}}} \quad [4.82]$$

The maximum mechanical work is then given as:

$$MAX_1(W) = \frac{1}{G_T-\Delta S_I}\left[G_T\left(T_{HS}\Delta S - T_{CS}(\Delta S + \Delta S_I)\right) - \Delta S(\Delta S + \Delta S_I)\left(\sqrt{T_{HS}} + \sqrt{T_{CS}}\right)^2\right] \quad [4.83]$$

A second sequential optimum is added to this first optimum, with respect to the variable ΔS. By derivation, it becomes:

$$\Delta S^* = \frac{1}{2}\left(G_T\frac{\sqrt{T_{HS}}-\sqrt{T_{CS}}}{\sqrt{T_{HS}}+\sqrt{T_{CS}}} - \Delta S_I\right) \quad [4.84]$$

and:

$$MAX_2(W) = \frac{1}{4(G_T-\Delta S_I)}\left[G_T^2\left(\sqrt{T_{HS}} - \sqrt{T_{CS}}\right)^2 - 2G_T\Delta S_I(T_{HS} + T_{CS}) + \Delta S_I^2\left(\sqrt{T_{HS}} + \sqrt{T_{CS}}\right)^2\right]$$

If $s_I = \frac{\Delta S_I}{G_T}$, the previous expression then becomes:

$$MAX_2(W) = \frac{G_T}{4\tau}\left[\left(\sqrt{T_{HS}} - \sqrt{T_{CS}}\right)^2 - s_I\left(\sqrt{T_{HS}} + \sqrt{T_{CS}}\right)^2\right] \quad [4.85]$$

The endoreversible maximum is standard and immediate.

Power optimization leads to:

$$\bar{W}(MAX_2(W)) = \frac{G_T}{4}\left[\left(\sqrt{T_{HS}} - \sqrt{T_{CS}}\right)^2 - \frac{C_I}{G_T\tau}\left(\sqrt{T_{HS}} + \sqrt{T_{CS}}\right)^2\right] \quad [4.86]$$

where C_I is a new concept that we call *action of entropy production (J·s/K)*, $\Delta S_I = \frac{C_I}{\tau}$.

Note that this mean power cancels out for $\tau \to \infty$, which corresponds to the known reversible thermostatic case, but also for a finite value τ'. So, the value $\tau^* \in [\tau' \,;\, +\infty]$ corresponds to an optimum.

$$\tau^* = \frac{2C_I}{G_T}\left(\frac{\sqrt{T_{HS}}+\sqrt{T_{CS}}}{\sqrt{T_{HS}}-\sqrt{T_{CS}}}\right)^2 \qquad [4.87]$$

We thus find for the maximum power:

$$MAX\left(\bar{W}\right) = \frac{G_T^2}{16C_I}\left(\frac{\sqrt{T_{HS}}-\sqrt{T_{CS}}}{\sqrt{T_{HS}}+\sqrt{T_{CS}}}\right)^2 \left(\sqrt{T_{HS}} - \sqrt{T_{CS}}\right)^2 \qquad [4.88]$$

4.8. Chapter conclusion

The thermodynamics of heat engines has experienced a revival over the past 50 years, although it has been studied for about a century and a half.

Here, we discuss the progression starting from the reversible Carnot cycle (1824), then from an endoreversible system with irreversible heat transfer at the source (Chambadal model) with thermostat source and (or) finite source, then the formally symmetric Curzon–Ahlborn model.

Our approaches complement this model published in 1975 by showing either a mechanical energy objective or a mechanical power objective.

Then, optimal state variables appear as:

– intensive variables T_H, T_C;

– extensive variables of finite physical dimensions $G_H, G_C, \Delta S$;

– the variable cycle duration τ.

This last variable accounts for the overall irreversibilities of the converter. According to the conditions imposed, we can observe whether or not an optimal period appears for the engine, the time leading to a maximum of average power, or an optimal distribution of the transformation times.

The interested reader will find supplements in recent publications. Research is continuing, not only on the characteristic case of the Carnot engine, but also on other engines, even on reverse cycle engines (see the books to follow on this same topic).

4.9. Appendix

$$MAX_1(W) = \varepsilon_H \dot{C}_H \tau \left(\sqrt{T_{HSi}} - \sqrt{T_0}\right)^2 - T_0 \frac{C_I}{\tau}$$

C_I, entropy generation of the converter.

$$\frac{\partial(MAX_1(W))}{\partial \tau} = \varepsilon_H \dot{C}_H \left(\sqrt{T_{HSi}} - \sqrt{T_0}\right)^2 + T_0 \frac{C_I}{\tau^2} > 0$$

$MAX_1(W)$ increases according to τ, regardless of C_I.

$$MAX_1\left(\bar{\dot{W}}\right) = \varepsilon_H \dot{C}_H \left(\sqrt{T_{HSi}} - \sqrt{T_0}\right)^2 - T_0 \frac{C_I}{\tau^2}$$

Therefore, $MAX_1\left(\bar{\dot{W}}\right)$ increases according to τ, regardless of C_I.

$$\frac{\partial\left(MAX_1(\bar{\dot{W}})\right)}{\partial \tau} = +2T_0 C_I \tau^{-3} > 0$$

But for an infinite energy source of according to τ.

For a finite source, $\varepsilon_H \dot{C}_H \tau$ is finite (G_H finite).

$$MAX_1(W) = G_H \left(\sqrt{T_{HSi}} - \sqrt{T_0}\right)^2 - T_0 \frac{C_I}{\tau}$$

Source power is optimal for:

$$\frac{\partial\left(MAX_1(\bar{\dot{W}})\right)}{\partial \tau} = -\frac{G_H\left(\sqrt{T_{HSi}}-\sqrt{T_0}\right)^2}{\tau^2} + \frac{2T_0 C_I}{\tau^3} = 0$$

With $G_H\left(\sqrt{T_{HSi}} - \sqrt{T_0}\right)^2 = W_{endo}$.

$$\frac{\partial\left(MAX_1(\bar{\dot{W}})\right)}{\partial \tau} = \left[-W_{endo} + \frac{2T_0 C_I}{\tau}\right]\frac{1}{\tau^2} = 0$$

$$\tau^* = \frac{2T_0 C_I}{W_{endo}}$$

$$MAX_1(W) = W_{endo} - \frac{T_0 C_I}{2T_0 C_I} W_{endo} = \frac{W_{endo}}{2}$$

$$MAX_1\left(\bar{\dot{W}}\right) = \frac{W_{endo}}{2}\frac{W_{endo}}{2T_0C_I} = \frac{W_{endo}^2}{4T_0C_I}$$

$$\eta(MAX_1(W)) = \eta\left(MAX_1\left(\bar{\dot{W}}\right)\right) = \frac{W_{endo}}{2Q_H}$$

With $Q_H = G_H(T_{HSi} - \sqrt{T_{HSi}T_0})$

Therefore:

$$\eta\left(MAX_1\left(\bar{\dot{W}}\right)\right) = \frac{1}{2}\left(1 - \sqrt{\frac{T_0}{T_{HSi}}}\right)$$

4.10. References

Andresen, B. (1983). Finite-Time Thermodynamics. Physics Laboratory II, University of Copenhagen.

Chambadal, P. (1957). *Les centrales nucléaires*. Armand Colin, Paris.

Curzon, F.L. and Ahlborn, B. (1975). Efficiency of a Carnot engine at maximum power output. *Am. J. Phys.*, 43, 22–24.

Feidt, M. (1987). *Thermodynamique et optimisation énergétique des systèmes et procédés*, 1st ed. Tech & Doc – Lavoisier, Paris.

Feidt, M. (1996). *Thermodynamique et optimisation énergétique des systèmes et procédés*, 2nd ed. Tech & Doc – Lavoisier, Paris.

Feidt, M. (2013). *Thermodynamique optimale en dimensions physiques finies*. Hermes-Lavoisier, Paris.

Feidt, M. (2014). *Génie Énergétique*, 2nd ed. Dunod, Paris.

Feidt, M. (2017). The history and perspective of efficiency at maximum power of the Carnot engines. *Entropy*, 19, 369.

Feidt, M. and Costea, M. (2019a). A new perspective on Carnot and Curzon-Ahlborn models of thermomecanical engine by transfer and production of entropy. *15th Joint European Thermodynamic Conference, JETC 2019*, May 22–24, Barcelona.

Feidt, M. and Costea, M. (2019b). Progress in Carnot and Chambadal modeling of thermomecanical engine by considering entropy production and heat transfer entropy. *Entropy*, 21(12), 1232.

Feidt, M. and Feidt, R. (2021). Du rôle fondamental de l'entropie de transfert de chaleur et de la production d'entropie dans la description et l'optimisation des moteurs thermomécaniques : le cas emblématique de la machine de Carnot. *Eur. Phys. J. Appl. Phys.*, 93, 60901.

Heywood, J.B. (1988). *Internal Combustion Engine Fundamentals*. McGraw-Hill Book Company, New York.

Petrescu, S., Costea, M., Feidt, M., Ganea, I., Boriaru, N. (2015). *Advanced Thermodynamics of Irreversible Processes with Finite Speed, and Finite Dimensions*. AGIR, Bucharest.

Vaudrey, A., Lanzetta, F., Feidt, M. (2014). H.B. Reitlinger and the efficiency at maximum power formula for heat engines. *J. Non-Equil. Thermodyn.*, 39(4), 199–203.

5

Circular Thermoeconomics: A Waste Cost Accounting Theory

Antonio VALERO-CAPILLA and César TORRES
CIRCE Institute, University of Zaragoza, Spain

The theory of circular thermoeconomics is presented to assess the physical cost of recycling waste in increasingly complex industrial processes. This is a crucial need since every time waste is recycled at a lower cost than its virgin alternative, we will save much sought-after natural resources. The theory allows us to generalize the idea of waste as any flow of useless mass and/or energy. It is simply the external irreversibility that generates entropy outside the system under analysis. Therefore, for an industry that is fully responsible for its waste, the true cost of waste should be the exergy of the resources used to produce, transfer and dispose of it within the system. To do this, it is necessary to identify an irreversibility carrier that is accountable for the accumulated entropy of the waste, from where it is formed to where it is disposed of.

The keyword *recycling* must be intrinsically related to efficiency and cost. It is used interchangeably by thermoeconomics, circular economy and industrial symbiosis. It allows the theory and its corollaries to be generalized since the only thing that changes is the type of system being studied. Among its conclusions, there is a *pinch point of recoverability* where total recycling is unattainable because a further increase in recycling does not correspond to the same efficiency increase. As a matter of fact, it fulfills the law of

Advances in Thermodynamics and Circular Thermoeconomics,
coordinated by Michael FEIDT and Antonio VALERO-CAPILLA.

diminishing returns and is a plea for the spiral economy instead of the common circular economy.

5.1. Introduction

Waste is inseparable from production and has always been considered an annoying and hidden part of it. Producing goods is now cheaper than ever since wasting and plundering natural resources is cheap. Capital is also affordable, and labor is gradually being replaced by automation, but robotics ignores how to deal with mountains of waste and restore nature[1].

We accept paying for clean air, purified water and hiding solid waste, even for planting trees. Still, we are not aware of the amount of waste we produce or its future effects through the depredation of resources. Ultimately, we are turning nature into a global dumping ground. The truth is that our society continues to increase its volume of waste, instead of decreasing it.

From a thermodynamic point of view, the generation of waste is not limited to solid or liquid materials, but also any unwanted emission of gas, sound, heat flow or radiation of any wavelength. They all increase the entropy of the environment, so there is no physical reason for their conceptual separation. In fact, it is not possible to live without producing waste, but as intelligent beings, we must assume that treatment and minimization have a cost that must be internalized in the production, the use of goods and their end-of-life. We must redefine efficiency, which is not only about making optimal use of resources, but also about taking responsibility for waste.

This radical vision of waste shows, for example, how the incineration of solid waste, which is considered socially acceptable – as long as there is strict control of polluting emissions such as volatile organic compounds or others – is not the best physical solution. The conversion of solid waste into useful heat and gaseous emissions increases the effects of climate change, as gases are diluted in the atmosphere. In contrast, solid waste takes up space and is a nuisance. It is a remarkable social trick that turns a local problem into a global one.

The problem of what to do with waste and emissions is, at heart, one of the biggest issues of the 21st century. There will not be raw materials for

1. This chapter is partially based on the paper *The Exergy Cost Revisited* (Torres and Valero 2021).

everything and everyone. If we continue like this, there may come a time when we have to choose between living off waste or destroying the planet. We must move from a linear economy of use and throw away to one in which circularity is central.

It is a costing issue, if we can get waste recycled at a lower cost and in conditions equivalent to its virgin alternative, we could reuse it. In a broad view, we should analyze both the costs of production and recovery, as well as all the waste induced in that process. This new waste will at least have abatement costs that will have to be calculated and added to the cost of the recovered waste. This process is by no means elementary. A systemic analysis that takes into account the thermodynamics of the process is needed.

To make the problem more difficult, the production of waste is always mixed. Let us take combustion as an example. Historians say that civilization was born with fire. Combustion is our most familiar way of obtaining heat, and all industrial activities use it directly or indirectly. However, to obtain heat, we produce gases and ashes. These gases are composed, among others, of CO_2, nitrogen oxides, volatile organic compounds, as well as water vapor, nitrogen and oxygen. The ashes have the inorganic matter from which the fuel is made plus unburned compounds. Both gases and ashes have reactivity and a long geological life ahead of them. They are neither pure nor ready to be economically reused in other processes.

The problem of entropy generation is presented here in its most perverse form. Separating calcium, magnesium, iron, boron, vanadium, phosphorus or uranium from the ashes of coal that have been burned a few moments ago requires much more energy than the electricity it has generated. Even the innocent act of washing our hands dilutes the soap and dirt, and pollutes the water we use, which will have to be purified later at a high physical cost, and certainly with additional waste.

Mixing does not require energy, and it is a spontaneous process. On the contrary, separating, de-mixing or de-alloying requires immense amounts of energy that make it physically and economically unviable in most cases. Moreover, what is worse, any of these industrial processes of separation produce more waste than those that try to separate. In other words, neither the *circular economy* (Ellen MacArthur Foundation 2013) nor *industrial symbiosis* (Graedel and Allenby 1995), nor of course *cradle-to-cradle* (McDonough and Braungart 2003) or *cleaner production* techniques (El-Haggar 2007), solves the problem of waste, although they obviously alleviate it. Whatever we do, entropy will grow, and it will grow all the more, the greater our ignorance.

At the theoretical level, it is necessary to connect economics with thermodynamics, not only to determine the rational cost of products, but also the cost of their associated waste. For this, it is required to investigate the causes of their formation in the thermodynamics of industrial processes. This requires a good physical accounting tool. A general theory to evaluate the costs of the process industry allows us to determine the impact, not only on CO_2, but on the entire waste production, as well as the backpack of critical raw materials and their depletion generated by the unbridled social demand. These costs must be, first and foremost, physical, i.e., they must account for measurable magnitudes not subject to the volatility of money and proportional to the physical deterioration they produce. The theory must be free of the price and subjective value of things on the economic side, and on the thermodynamic side, look for functions that reflect the loss of potential of physical currents (understanding by potential the capacity to do useful work). Therefore, entropy must be incorporated in the analysis as a rigorous measure of the constant loss of quality suffered by the real processes of production of goods.

In short, a general theory is needed to evaluate the real cost of waste and how much it costs to recycle it in increasingly complex industrial processes, either before it reaches the exterior or in reconditioning processes so that it can later be used as a secondary resource. In this chapter, we want to show how thermoeconomics has precise answers.

Thermoeconomics (see, for example, Valero and Torres (2009)) may be defined as an application of the general systems theory (Bertalanffy 1968) that thermodynamically describes a set of articulated subsystems interconnected by energy/mass streams to seek the amount of resources (physical and/or economic costs) needed to produce each and every energy interacting flow. The subsystems can range from simple physicochemical processes, equipment and components of a factory, installations belonging to an industrial complex or even the processes that a product undergoes throughout its lifecycle and recycling. When the system refers to a structured industrial site, we will refer to industrial symbiosis. When the product recycling process is considered, we will refer to the generic name of circular economy.

The task of reducing internal irreversibilities of a system is called energy efficiency. In contrast, the task of reducing external irreversibilities, also called waste, is referred to as *spiral economy* (Valero and Valero-Delgado 2014; Ashby et al. 2019), rather than circular economy. Meanwhile, the study of assessing the cost of perturbations of the system is called thermoeconomic diagnosis. Besides that, system design is the skill of assembling a set of subsystems to produce a cost-effective product subject to several constraints. In comparison, eco-efficiency is the art of designing products to facilitate their recycling at the end of their life. All these techniques can be studied from the perspective of thermoeconomics, as briefly described in this chapter.

This theory will serve to answer questions such as Is a "Zero Waste" society possible? What does "Zero Waste" really mean? What are the limits of circular economy from a thermodynamic point of view? How can the cost of recycling waste be calculated? What physical properties characterize waste to be recycled at low costs? How do we anticipate the costs of recycling products-to-be-waste? What does the internalization of costs of waste mean in terms of thermoeconomics? How do we assess the true costs of cascading-waste-into-feedstock techniques of industrial symbiosis? and so on.

This chapter introduces the concept of generalized waste as an undesirable companion to all production and even as its final destination. It studies to what extent the exergy of all waste is an opportunity to improve the efficiency of processes, and what the very limits of recovery and its physical costs are. It defines a universal theoretical basis that would serve as a perspective tool for the new technologies associated with waste recovery.

5.1.1. *Toward a general theory of process efficiency*

As it is well known, exergy measures the amount of useful energy that can be extracted from an energy flow when faced with surroundings that are taken as the reference environment. In any stationary process, the balance of input exergy minus output exergy is equal to the ambient temperature multiplied by the entropy generated within the process (Gouy–Stodola's law). This product is known as the irreversibility of the process and measures the internal degradation of energy quality, allowing it to be counted in standard energy units.

Consequently, the efficiency expressed in terms of exergy is more rational than the usual energy efficiency. A clear example can be seen in spontaneous processes such as heat transfer, pressure loss or dilution, in which the quantity of energy is not altered, but its quality. It leads to a loss of exergetic potential, and therefore a generation of entropy. The only way to save energy is to avoid irreversible processes whose magnitude is assessed by exergy balances. As long as energy is neither created nor destroyed, exergy is continuously and spontaneously destroyed. Therefore, it makes perfect sense to propose a general theory of exergy saving instead of energy saving.

At the annual ASME Conference in 1986, we presented the *general theory of exergy saving*, (Valero et al. 1986) later called the *exergy cost theory* (Lozano and Valero 1993). It is based on three ideas, the first being that the loss of process efficiency must be measured using exergy and the second that it can be identified in all the components of a thermal system, using a resource, called fuel and a product, which defines its efficiencies (Tsatsaronis and Winhold 1985). Among its messages is the fact that the real physical cost of the products is found from the sum of the irreversibilities assigned in their formation process plus the exergy contained in the product. In this way, the concept of cost, coming from the economy, was connected to the concept of irreversibility, beyond the identifiable energy losses. Second, in complex energy systems, an improvement in the efficiency of some component of the system does not necessarily imply a saving of resources. This is because the impact of any local improvement can negatively affect the rest or some of its components (rebound effect), so the energy-saving must be structurally assessed. Third, it was demonstrated that a set of efficiencies defined in a productive structure induces one and only one set of costs of the flows that interact with it and vice versa. In other words, the concepts of efficiency and cost are inseparably linked. Thus, the structural theory of thermoeconomics (Valero et al. 1993) showed that, for each production plant, the set of efficiencies of its processes, called the primal set, corresponds to a unique dual set, formed by the production costs of its processes. When industrial plants are designed with a productive purpose, the set of efficiencies of their components will also be linked to that purpose, as well as to the set of their production costs.

Although the concept of exergetic cost coincides with that of cumulative exergy consumption proposed by Szargut et al. (1988), the idea of exergy cost reinforces the connection between thermodynamics and economics. Indeed, its theoretical framework is mathematically isomorphic to Leontief's input–output theory (Leontief 1970), which was developed to analyze sectors of a country's economy. However, there is a fundamental distinction, the irreversibility is postulated as the physical origin of the concept of cost. While irreversibility is measured in physical units (kWh), money is the measure of economic cost. Moreover, both are distinguished by their object of analysis, since thermoeconomics is concentrated on the physical chemistry of industrial production processes. On the other hand, in the problems of linear optimization, the Lagrange multipliers or shadow costs are the weight of the structural restrictions that the optimum must fulfill. The structural theory shows that exergetic costs coincide with Lagrange multipliers in the optimization of the cost of a product constrained to a given production structure (Valero et al. 1989).

This set of reasons is more than enough to maintain the idea of exergetic cost. On the other hand, thermoeconomics is a physical theory based on the principles of thermodynamics and transfer sciences that owe their conceptual framework to economic theory. It is a science in transition between the measurable objectivity of physics and the subjectivity of the productive purpose. These reasons avail that the exergetic cost theory (ECT) and structural theory (ST) can be considered as the mathematical basis of thermoeconomics.

5.1.2. *Basic concepts*

If irreversibility represents the *useful* energy lost or destroyed in all physical processes, it can be used to pinpoint the true inefficiencies of industrial processes. Since all processes in an actual plant are not reversible, exergy is destroyed, and some natural resources are consumed and lost forever, which involves a cost in physical and economic terms. The more irreversible a process is, the more natural resources are consumed.

The exergy balance accounts for the degradation of the exergy. The input exergy of a process will always be greater than the exergy output:

exergy input – exergy output = irreversibilities > 0

This expression only keeps in mind the irreversibilities of the process. But, the purpose of a process is set by means of the definition of its efficiency. This is to say that there is an implicit classification of the streams crossing the boundary of the system: the flows that are the production objective, the resources required to carry out the production and those that are wasted. This information is not implicit in the second law and is the most important conceptual leap separating and at the same time uniting physics with economics. The following equation goes beyond the second law:

Resources (F) – products (P) = Irreversibilities (I) + wastes (R) > 0

Besides that, this equation allows the concept of efficiency in productive units to be defined

Efficiency = Products (P)/Resources (F)

This definition is older than thermodynamics itself and measures the quality of a process. The desire to produce a certain product is external to the system and must be defined beforehand. Once this has been done, the design of the system and its functional structure will fit the aim of using available resources (capital, raw material, manpower, etc.). Every definition of efficiency demands a comparison of the product obtained with the resources needed to obtain it. Its inverse is:

Unit consumption = Resources (F)/Product (P)

This expression is also a definition of the average unit cost when resources refer to the overall plant instead of individual processes. This concept is the key to thermoeconomics.

Thus, thermoeconomics may assess the cost of consumed resources in terms of system irreversibilities as well as in monetary accounts. They help to point out how resources may be used more effectively in order to save them. Money costs express the economic effect of inefficiencies and are used to improve the cost effectiveness of production processes. Assessing the cost of the flows and processes in a plant helps to understand the process of cost formation, from the input resources to the final products.

However, thermoeconomics developed so far has focused mainly on the production of goods, mainly energy commodities, leaving aside the production of waste, although both are inseparable. From a thermodynamic point of view, a waste is a useless physical current that crosses the boundaries of a system. But all waste has exergy, either thermal (heat), mechanical (noise) or chemical (reactivity), and different concentrations cause pollution in the environment. This exergy, sooner or later, ends up dissipating into the environment, causing external irreversibilities. That is why the internal irreversibilities already accounted for in the cost of production must be added to the external irreversibilities that cause their waste.

In other words, the theory of exergetic cost and structural theory must be extended to include the formation of waste as a fundamental part of the calculation of production costs. Therefore, this is the subject of this chapter.

EXAMPLE.– In order to illustrate the methodology introduced in this chapter, a simple model of a gas turbine cycle with co-generation (henceforth called TGAS) is used as a case study throughout this chapter (see Figure 5.1). It uses natural gas (flow #1) to produces 10 MW of electric power (flow #7) and 7.6 kg/s of saturated steam at 20 bar of pressure (flow #8) from freshwater (flow #8a), partially recovering the residual heat of gases (flow #4). Finally, the waste combustion gases are expelled through a stack (flow #9). The combustor has 2% of heat losses, and the isentropic efficiency of the turbine and compressor is 87%. This flowsheet also indicates the system boundaries and the aggregation level. The thermodynamic properties of the plant are shown in Table 5.1. To simplify the model, flows #0 and #8a are removed from further analysis because their exergy values are equal to zero. We have not considered the pump required to increase the water pressure from 1 to 20 bar because the exergy value of the work flow is negligible (1.8 kW).

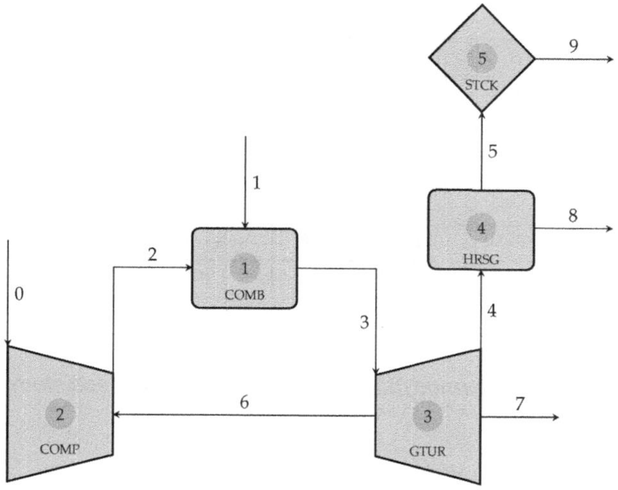

Figure 5.1. *Flowsheet of TGAS plant*

5.2. The exergy cost theory

From a theoretical viewpoint, an energy system is defined beforehand as a set of subsystems or thermodynamic processes linked to each other and to the environment by another set of mass, heat and work flows, quantified by its exergies or in general its thermodynamic properties. In essence, it constitutes a flow sheet, as shown in Figure 5.1. This flowsheet also indicates the system boundaries and the aggregation level.

Industrial installations have a productive purpose. Their objective is to obtain one or several products by processing external resources. For each process of the system, it is necessary to identify the flows that constitute their product, and the flows required to obtain them, called fuel. Energy flows (e.g. heat or work) appear at the process inlet as fuel stream, or at the process outlet as product stream. When working with mass streams, it is appropriate to operate with exergy differences associated with each energy transfer between the inlets and the outlets. The output flows that are not part of the production process are considered as unspent or *recoverable fuel*.

Nr	Flow	m (kg/s)	T (C)	p (bar)	h (kJ/kg)	s (kJ/kg K)	B (kW)
0	ENV	37.53	20	1.013	0	0	0
1	NG	0.7324	20		50,000	0	36,620
2	B2	37.53	291.4	8	272.4	0.0648	9,512
3	B3	38.26	1050	7.84	1,205	1.170	32,990
4	B4	38.26	598.2	1.05	676.5	1.264	11,707
5	B5	38.26	137.4	1.013	137.3	0.394	835.3
6	WC						10,225
7	WN						10,000
8a	ENV	7.6	20	1.013	83.9	0.296	0
8	QV	7.6	212.4	20	2,799	6.340	7,168
9	QG	38.26	137.4	1.013	137.3	0.394	835.3

Table 5.1. *Thermodynamic properties of TGAS case study*

A fuel stream group will consist of:

– one or several inflows that provide exergy to the process;

– input and output flows that enter into the process and leave it after some transfer of exergy to the process.

A product stream group will consist of:

– one or several outflows produced by the process;

– output and input flows that enter into the process and leave it after some exergy increase.

The exergetic efficiency of a unit component u is defined as the ratio between the exergy obtained or product P_u and the exergy supplied or fuel F_u. It measures how reversible a process is. The inverse value is the exergy unit consumption, that is, the amount of resources required per unit of the product obtained:

$$k_u = \frac{F_u}{P_u} \geq 1 \qquad [5.1]$$

The second law states that the exergy of the fuels is bigger than the exergy of the products, and the difference is the irreversibility of the process, both internal and external.

$$F_u - P_u = I_u \geq 0 \qquad [5.2]$$

Consider a generic component, as shown in Figure 5.2, whose unit exergy consumption is defined as:

$$k = \frac{B_1 + (B_2 - B_3)}{B_6 + (B_5 - B_4)}$$

The objective of the process is to produce flow B_6 and increase the exergy of flow 4 to B_5. For this purpose, two fuel streams are defined: the flow B_1 and the exergy transferred by the difference $B_2 - B_3$. Flows #1, #2 and #3 are associated with the fuel streams, whereas flows #4, #5 and #6 are associated with the product streams.

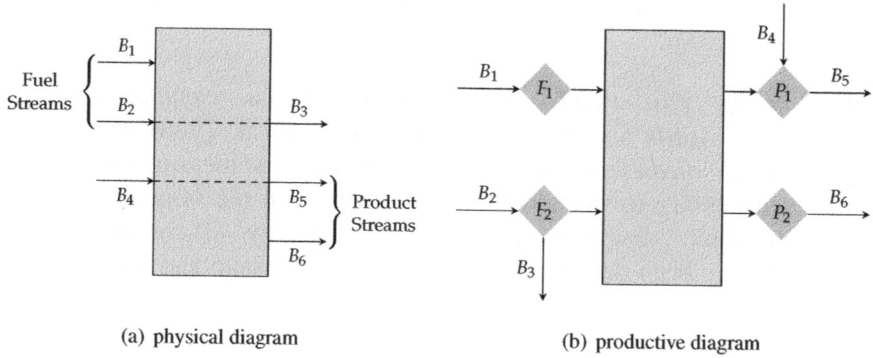

Figure 5.2. *Physical and productive diagrams of a generic component*

In the productive or logical diagram, the stream groups are represented by diamonds. They are not physical processes, but logical connectors between flows and processes, which contain information about the processes efficiency.

EXAMPLE.– In the TGAS example, the exergy of process steam and the net power are the products of the global system, and the exergy provided by the natural gas is the fuel. The purpose of the compressor is to increase the pressure (and the exergy) of the air from flow #0 to flow #1 using the mechanical work produced by the turbine. The aim of the combustor is to increase the exergy of the air flow B_1 that exits from the compressor to B_2, and the exergy of natural gas B_5 is consumed as fuel. In the case of the turbine, the aim is to obtain mechanical energy. Therefore, the product is the exergy employed to drive the compressor B_6 and the net power of the plant B_7. The exergy provided by the gas expanded in the turbine $B_2 - B_3$ is the fuel. Finally, the purpose of the HRSG is to produce process steam, flow #8 using part of the exergy of the exhausted gases $E_3 - E_4$. The unrecoverable gases, flow #4, are expelled in the stack, flow #9. Table 5.2 defines the efficiency of the TGAS processes.

Nr	Key	Description	Fuel	Product
1	COMB	Combustor	B_1	$B_3 - B_2$
2	COMP	Compressor	B_6	B_2
3	GTRB	Turbine	$B_3 - B_4$	$B_6 + B_7$
4	HRSG	HRSG	$B_4 - B_5$	B_8
5	STCK	Stack	B_5	B_9

Table 5.2. *Process description and efficiency definition of TGAS plant*

The *exergy cost* (kWh), otherwise known as *cumulative exergy consumption* (Morris and Szargut 1986; Szargut et al. 1988) or presently adopted as *embodied exergy* according to some authors, measures the amount of exergy necessary to manufacture a product when the boundaries of the production plant, disaggregation level and exergy efficiency of each component have been defined (Valero et al. 1986; Lozano and Valero 1993). Herein, the exergy cost of a stream with exergy B is denoted with an asterisk, B^*, and its unit exergy cost is expressed as follows: $k^* \equiv B^*/B$.

The process of costs allocation in productive systems requires a set of rules that were first presented in Valero et al. (1986):

– *Resources Rule*: the exergy cost of the flows entering a system should be provided as input data. In the absence of further information about its cost, it can be assumed to be equal to the exergy.

– *Cost Conservation Rule*: for each and every system component, the exergy cost of output flows is equal to the exergy cost of the input flows.

– *Co-products Rule*: it is also known as Rule P. All products of the same quality at the output of a system component have the same unit exergy cost.

– *Recoverable Fuel Rule*: it is also known as Rule F. The unit exergy cost of the outflows that are not part of the main product, but considered as recoverable or unspent fuel, is the same to that of the input fuel.

– *Wastes Internalization Rule*: the exergy cost of a waste stream leaving the boundaries of the system is allocated null and endorses this cost on the final product(s). In fact, its exergy cost, formed by adding the fuel required to produce it plus the fuel used to carrying and dispose of it, must be allocated to the components that have generated it.

In summary, the (average) exergy cost is a conservative property and the production cost must be allocated to what we want to produce. Therefore, when a process has multiple outputs, rules F and P provide a solution to cost allocation, based on the efficiency definition of the processes. The last rule of *waste internalization* was added to the previous ones (Torres et al. 2008).

Let us see an example of the application of these rules. Suppose first, a medium pressure steam turbine, as shown in Figure 5.3, where the entering steam F_1 has 30 MW of exergy, it delivers a shaft work of $P = 10$ MW and the exit steam F_2 still has 19 MW of exergy. Then, the exergy expended to produce P is $F_1 - F_2 = 11$ MW, and its efficiency $\eta = P/(F_1 - F_2) = 0.909$.

If the turbine's objective is to produce power, the full cost of that product should be allocated to the shaft work. Accordingly, the rule F means that the unit exergy cost of the flowing steam remains unaltered. Suppose that the amount of exergy expended to produce the entering steam is 2.5 times greater than its exergy content, that is, $k_1^* = 2.5$ MW/MW. Then the rule F states that the unit exergy cost of the exit steam, F_2, is that of the entering steam, F_1, or $k_2^* = k_1^* = 2.5$. Consequently, the amount of exergy needed to produce the shaft work will increase to $k_3^* = k_1^*/\eta = 2.75$ MW/MW.

To explain the *co-products* rule, consider an electric generator added to a distributor, as shown in Figure 5.4. It delivers two currents, 9 MW of net energy output and 0.5 MW for ancillary systems, at the expenditure of 10 MW of shaft work. Its efficiency becomes $\eta = (P_1 + P_2)/F = 0.95$. As both outlet streams are simultaneously required, the cost of its production must be allocated proportionally to both currents. Therefore, the *Rule P* states that the exergy cost per unit of exergy must be equal for the two currents. If the unit exergy cost of the shaft work was $k_1^* = 2.75$ MW/MW, then the unit costs of the output electric streams become $k_2^* = k_3^* = k_1^*/\eta = 2.895$ MW/MW. The reader may find many examples of the application of these rules to a wide-ranging class of thermal systems in Bejan et al. (1996).

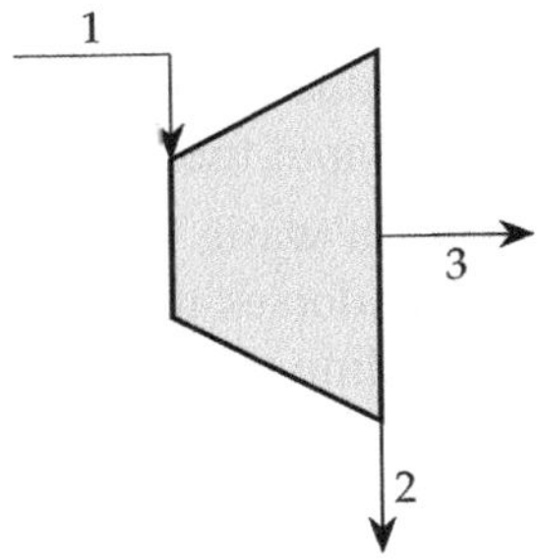

$$k = \frac{1}{\eta} = \frac{F}{P} = \frac{B_1 - B_2}{B_3}$$

and

$$k_2^* = k_1^*$$

$$k_3^* = k\, k_1^*$$

Figure 5.3. *Conventional efficiency definition of a turbine, as an example of the recoverable fuel rule application*

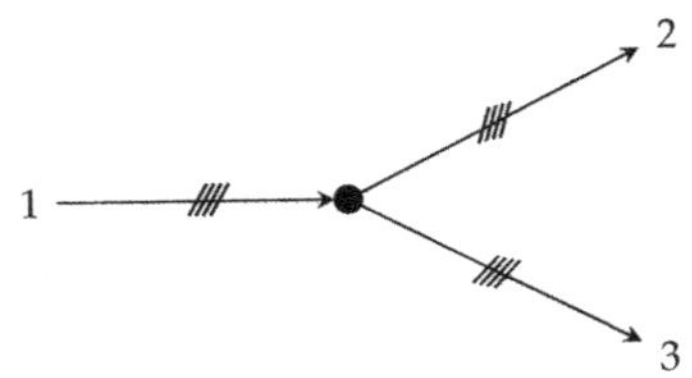

$$k = \frac{1}{\eta} = \frac{F}{P} = \frac{B_1}{B_2 + B_3}$$

and

$$k_2^* = k_3^* = k\, k_1^*$$

Figure 5.4. *Generic electric power distribution system as an example of application of co-products rule*

As it can be seen in this simple example, these rules link the concepts of the unit exergy cost, k^*, of the streams with the efficiency definition of the system's components, $\eta = 1/k$, one-to-one, where k is named as the unit exergy consumption of the component. In other words, the cost of the final product increases as irreversibilities accumulate. If these increase, the efficiency of the components decreases and the costs increase accordingly. Thus, there is a direct connection between efficiency, cost and irreversibility, when exergy is used as a measure of the energy flows of thermal systems. This is the essential message of the theory of exergy cost, which has been extensively used to assess the costs of many thermal systems.

Consider now a simple system, as shown in Figure 5.5.

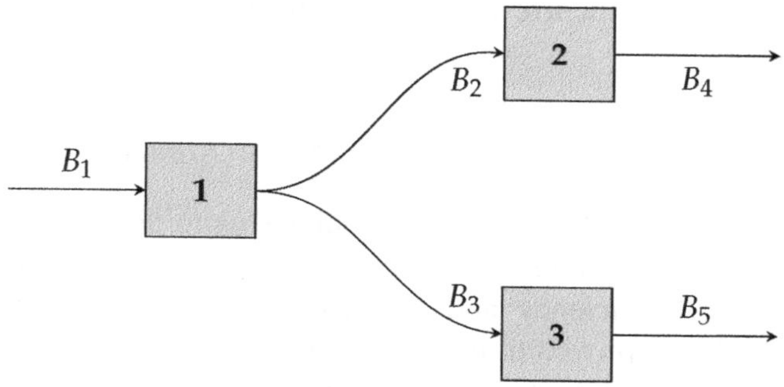

Figure 5.5. *Generic system with several product bifurcations*

According to the exergy cost assessment rules, the cost of flows will be:

$$B_1^* = B_1$$
$$B_2^* = y_2\, B_1 = B_2 + y_2\, I_1$$
$$B_3^* = y_3\, B_1 = B_3 + y_3\, I_1$$
$$B_4^* = y_2\, B_1 = B_4 + y_2\, I_1 + I_2$$
$$B_5^* = y_3\, B_1 = B_5 + y_3\, I_1 + I_3$$

where y_2 and y_3 are the exergy distribution ratios, defined as:

$$y_2 = \frac{B_2}{B_2 + B_3} \qquad y_3 = \frac{B_3}{B_3 + B_3}$$

To obtain the relationship between cost and irreversibility, consider the exergy balance of the processes: $B_1 = B_2 + B_3 + I_1$, $B_2 = B_4 + I_2$ and $B_3 = B_5 + I_3$.

The exergy cost accounts for the resources required to obtain a stream or flow of products. Furthermore, as shown in this example, the exergy cost of a flow also accounts for all the irreversibilities generated to obtain it. We have also seen that the exergy cost of any flow is always higher than its exergy content, and this is because the processes are not reversible. There is a direct relationship between the costs and the irreversibilities of the processes (Torres and Valero 2021), which could be written as:

$$B_i^* = B_i + \sum_{i=1}^{n} \pi_{ij}\, I_j \qquad [5.3]$$

where π_{ij} is the part of the irreversibility of process j that has been generated to obtain stream i.

5.2.1. *The principle of non-equivalence of local irreversibilities*

The exergy balance of an installation allows us to allocate and calculate irreversibilities in the production process and identify the equipment that affects the overall efficiency and the reasons thereof. This information, although helpful, has proved to not be enough. In practice, when attempting to achieve energy savings in an installation, we must consider that not all irreversibilities can be avoided. The potential exergy saving is limited by technical and economic constraints. Thus, the possibilities for exergy savings, which is called technical exergy saving, are always lower than the theoretical limit of thermodynamic exergy losses.

Therefore, the additional fuel consumption can be expressed as the difference between the resources consumption of the plant between operation (x) and reference or design conditions (x^0), with the same production objectives, that is, with the same total plant product $\Delta P_T = 0$.

$$\Delta F_T = F_T(x) - F_T(x^0)$$

It can be broken up into the sum of the irreversibilities of each component:

$$\Delta F_T = \Delta I_T + \Delta P_T = \sum_{j=1}^{n} \Delta I_j + \Delta P_T \qquad [5.4]$$

However, the local exergy savings achieved in the different units or processes of a plant are not equivalent. The same decrease in the local irreversibility of two different components generally results in different variations of the total energy consumption of the plant.

Using a thermodynamic model simulator or a plant performance test, we can obtain its exergy flow values and the efficiency of each individual process. From these operating data, unit exergy consumptions and other thermo-economic variables can be calculated for the operating conditions. These values can be compared with the corresponding plant design values. In general, during operation, several plant components suffer a degradation of their behavior with respect to the design conditions and, as a consequence, an increase of their irreversibility, given by:

$$\Delta I_j = \Delta k_j \, P_j^0 + (k_j - 1)\Delta P_j \qquad [5.5]$$

where P_j^0 is the production in the reference conditions $P_j(x^0)$. From the above expression, we can distinguish two types of irreversibilities:

– *Malfunction*: this is the irreversibility produced by an increase in the unit consumption of the component itself or endogenous irreversibility.

$$MF_j = \Delta k_j P_j^0 \quad [5.6]$$

– *Dysfunction*: this is the exogenous irreversibility induced in a component by the malfunction of other subsystems, which forces it to consume more local resources to obtain the additional production required by the other components.

$$DF_j = (k_j - 1)\Delta P_j \quad [5.7]$$

The degradation of one component forces the other components to adapt their behavior in order to maintain their production conditions and, consequently, their irreversibilities are modified. Figure 5.6 shows how an increase in the unit consumption of a component increases not only the irreversibilities of that component, but also those of the previous component. A malfunction $P_j \Delta k_j$ in a system's component, not only causes an increase on irreversibility on this process, but also in the rest of processes.

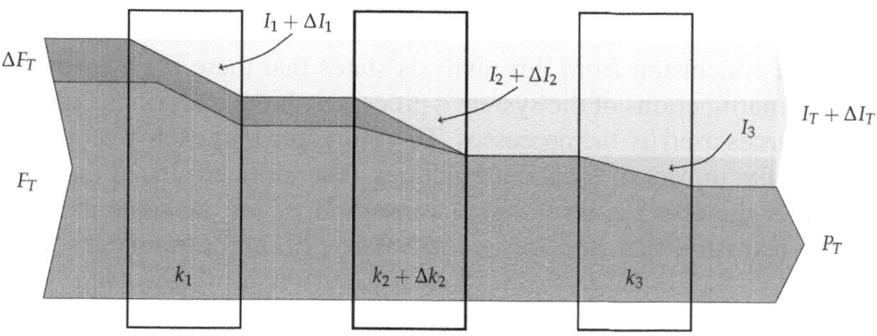

Figure 5.6. *Effect of a local malfunction in a generic system. For a color version of this figure, see www.iste.co.uk/feidt/advances.zip*

Consider a sequential system, as shown in Figure 5.6, composed of n processes. The total fuel is given as:

$$F_T = k_1 \cdots k_i \cdots k_n \, P_T,$$

and suppose that the variation of the total fuel is due only to the variation in the exergy consumption of the ith unit given by Δk_i, then the variation of resources entering the system could be written as:

$$\Delta F_T = k_1 \cdots k_{i-1} \Delta k_i k_{i+1} \cdots k_n P_T = k^*_{F,i} \Delta k_i P_i$$

The theory of exergy cost (Lozano and Valero 1993) defines the fuel impact as the effect on total, resources consumption, which can be achieved if the efficiency of a process, or equivalently its unit consumption, is improved. It will be assessed by multiplying its local malfunction $P_i \, \Delta k_i$ by the unit exergy cost of the fuel consumed by this process.

$$\mathcal{A}_{T,i} \equiv k^*_{F,i} \, P_i \, \Delta k_i = F^*_i \% \Delta k_i \qquad [5.8]$$

where $\%\Delta k_i \equiv \Delta k_i / k_i$ is the relative change of unit consumption. The fuel impact [5.8] states that the location of the irreversibility does not coincide with the location of the causes that provoke these losses. When a system's process degrades, we will need more local resources to obtain the required production. Moreover, the increase in the unit exergy consumption of a process causes the increase of its unit production cost and consequently the increase of the unit production costs of all processes downstream. The fuel impact formula expresses the non-equivalence of irreversibilities in different system processes, introduced by Beyer (Kotas 1985).

The first conclusion from this analysis states that there is no equivalence among the malfunctions of the system's processes. It depends on the unit costs of the resources used by the processes. Therefore, the unit exergy cost $k^*_{F,i}$ is a measure of the impact of local malfunctions. The more advanced the process is, the larger the effect is on resource consumption, for the same increase in the local irreversibilities. See Torres and Valero (2012) for more details.

5.2.2. *The cost formation process of waste*

An essential question for cost assessment is how internal and external irreversibilities manifest themselves in a plant and how they determine their layout. An internal irreversibility, due to a useless loss of pressure, temperature or chemical potential, can be located in any component of a plant, but its effects can appear anywhere as useless heat. This is the case when the location of a cause does not match the location of its effects.

Consider a Rankine cycle power plant in which only internal irreversibilities would occur. We know that for each unit and for the plant in general, $F_u - P_u = I_u = T_0 S_{g,u}$. Nevertheless, F and P have an energy content, H_F and H_P. Then, the first law also states for each and every unit as well as the overall plant: $H_{F,u} - H_{P,u} = Q_u$.

However, most of the processes are designed to have good thermal and acoustic insulation. Therefore, Q_u is not significant in terms of the overall heat released at the overall plant, $\sum Q_u$. In fact, it is observed that for each unit, Q_u is released and an irreversibility I_u is generated, but they are not related one by one, except at the level of the entire plant.

It is worth asking now, where does the heat released by each and every one of the units appear? Remarkably, the heat released is detectable, while irreversibility is a conceptual construct with no local manifestation. So where does such a heat release originate from a second law point of view? How do we understand this fact? How do we reconcile the first and second laws? The answer is obvious: at the level of the complete plant, it is true that: $S_{gen,total} = \sum Q_u/T_0$.

As it is well known, most of the heat released in a power plant is rejected in the boiler and condenser-cooling tower subsystem. Or in other words,

$$F_T - P_T = \sum Q_u = \sum I_u,$$

and note that at the whole power plant H_F nearly coincides with its exergy F and H_P is equal to P.

This an important idea, even if it seems evident. F_u and P_u are located in each and every component, however, released heats appear anywhere on the plant. In other words, F_u, P_u and I_u are located in each component, while each and every Q_u says little of the actual losses. This is a strong argument for using exergy in thermoeconomics. Using the second law allows the true causes of degradation to be located in the units where they appear. Therefore, the casualization of inefficiencies is better located.

We cannot pinpoint the causes of inefficiencies simply by looking at the different heat releases, even if local irreversibilities are eventually dispersed as heat fluxes to the environment. Besides that, it is not easy to identify the origin of each degraded heat among the different units in a plant. In any case, this unwanted heat must be eliminated, and additional dissipative equipment is needed to achieve this. This is the case of a condenser in a power plant. In other words, internal irreversibilities, sooner or later, generate waste. However, they need a mass medium to manifest. Let us call this medium an *irreversibility carrier*.

A waste stream has exergy, whether thermal, mechanical or chemical. In addition, it has its cost of formation, transference, partial recovery and disposal that must be identified. In parallel with the product cost formation process, there is a waste cost formation process. The irreversibility carrier concept allows us to identify the origin, path and disposal of any residual stream of the plant regardless of whether it is a mass stream or not. These flows still have exergy that the system does not recover. Rather, additional resources have to be spent to partially get rid of them in *dissipative units*. This is the case for the stack and its auxiliary systems for combustion gases. These units avoid a greater impact on the environment of waste streams and constitute a technological response to the necessary internalization of the costs of waste produced by industrial processes.

By contrast of internal irreversibilities, in the production of mass waste, waste is itself a carrier. It is easy to find the origin of waste production and also to follow such stream to the dissipative unit. In terms of mass, the industry produces more waste than commodities, and, worse, more types of residual streams than products. For example, in a chemical reaction such as combustion, we are interested in the heat released rather than the products of the reaction. Therefore, CO_2 and water, plus other unwanted side reaction products such as CO, NOx, SOx, unburned, ash and slag, become waste for disposal. Even in the case of reactors in industrial chemistry, metallurgy and all manufacturing processes in which a reaction product is obtained, the types of unwanted waste are always present. With these basic ideas, we can see how thermoeconomics can assess the costs of waste associated with internal and external irreversibilities. This process is found by following its irreversibility carrier, which identifies its origin, transfer, partial recovery and disposal. At the end of the analysis, this cost must be added to the final cost of the plant production. This process is dubbed as *cost internalization*.

Sometimes waste flows are caused by technological limitations imposed by the equipment and processes of the plant. However, it is unfortunately common to overlook the external effects of waste. This practice is called cost outsourcing or cost externalization, that is, letting society or the environment assume the impact of plant waste. This is the case for greenhouse gases produced by combustion processes. Therefore, the techniques of waste allocation or costs internalization will increasingly become a common practice in the near future. Waste flows still have exergy to be recovered or to get rid of. In any case, we need to spend additional resources, thus increasing the system's irreversibility. In accounting terms, if waste streams cross the boundary limits of the system, they can be considered as external irreversibilities. However, appropriate allocation needs to locate which internal process causes them.

For instance, the stack of TGAS plant does not cause the exiting gases. This waste stream actually has at least three types of residues: rejected heat, CO_2 and NOx (plus water, oxygen, nitrogen and other components of the incoming air, not considered waste. Also, in a gas turbine, we are unlikely to find CO and unburned fuel, but no strong noise, which is also an unwanted flow). Carbon dioxide and water are produced by the combustion of carbonated fuel, while NOx and excess heat are produced in the combustion chamber. You can compare this combustion chamber to a perfect electrochemical fuel cell. In such a case, only CO_2 and water would be produced.

Therefore, the cost of getting rid of such gases is always associated with the type of equipment and/or fuel used. In general, all the productive components involved in the formation of waste contribute to the formation of the cost of the waste. In short, non-material waste streams like noise, that is, excess pressure that causes vibrations, or excess temperature that turns into rejected heat are internal irreversibilities that eventually turn into waste heat. They need material carriers. However, these carriers easily blur their causal origin. The local entropy generation is the measure of the additional carrying load of the irreversibility carrier.

The remaining waste, CO_2 and NOx are in fact material streams that need *external* dissipative units to get rid of them. Carbon dioxide requires carbon capture and storage techniques, while NOx requires special combustion chambers to prevent its formation. These abatement processes require additional fuel and investments whose costs must be added to the plant product(s).

They are true streams of waste that have an origin, an itinerary and an end. Therefore, it is easy to identify their cost formation process as well as their logical disaggregation. Along the way, they add costs. First that of its formation exergy in the chemical reaction, then that of conduction and eventual separation, and finally their abatement costs. We call *external irreversibilities* to all irreversibilities caused by their treatment and abatement processes, as explained before. Similar to the definition of fuel and product, that of waste also depends on the analyst and ultimately on society. In fact, nature does not produce waste because it recycles everything. As long as waste has exergy, it has the capacity to be recovered. That is, a material stream can be recovered as long as it differs from its reference environment.

The distinction between internal and external irreversibilities will also depend on the analyst. However, the cost of any malfunction in the plant must have a single value regardless of the classification of irreversibilities.

Therefore, for a proper allocation of wastes costs, we must follow their formation process and assess the cost of their formation and disposal to the processes in which they are produced, according to the ideas of the *irreversibilty carrier* and the *polluter pays principle*. In parallel to the cost formation process of the products, there is a cost formation process of wastes.

5.2.3. *Waste flows cost allocation*

Let $\mathcal{V} = \{u_0, u_1, \ldots, u_n\}$ be the set of the n components or processes whose constitute the system, for some defined boundaries and aggregation level. The component u_0 represents the environment. Define $\mathcal{B} = \{b_1, \ldots, b_m\}$ as the set of m flows that interchange exergy between the processes, and $\mathcal{R} \subset \mathcal{B}$ as the subset of waste flows.

From the viewpoint of processes, we must distinguish among productive components, whose purpose is to make useful products, and dissipative components, that do not generate any final product, but are responsible for disposing of the wastes created during the production. For convenience, each waste flow is defined as the product of a dissipative unit, so the waste cost reallocation could be made from dissipative units to productive components.

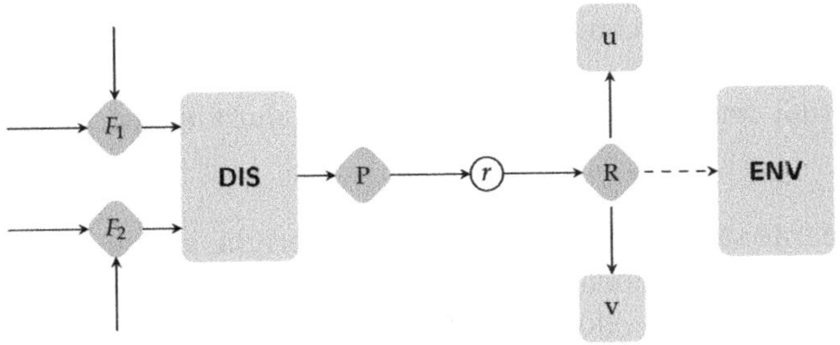

Figure 5.7. *Waste allocation scheme in the productive structure of a system*

Figure 5.7 shows the cost allocation scheme for dissipative processes and waste flows streams in the productive structure. Flows to be disposed of enter as a fuel of the dissipative unit (DIS) together with the resources required to eliminate them.

The product of the dissipative unit is the removed exergy, flow (r). This flow is connected with the environment through a logical block (diamond shape), named the *waste stream group* that redistributes its cost among the productive units (u) and (v) that originate it. Meanwhile, the flow cost leaving the system (entering the environment) is zero. In other words, waste streams are treated as if they are external irreversibilities of the productive units that generated them.

In the case of the TGAS example, the flow #4 enters on the fuel group F_1, and the F_2 group is not used in this case, but it could represent the work and electricity flows of the equipment required to extract the gases (forced or induced draft fans). The purpose of the dissipative unit is to eliminate the exergy of output gases B_9. The waste stream group R is a logical node, whose function is to redistribute the cost of gases B_9^* among the processes having contributed to the generation of the waste flows, in this case, the combustion process. Therefore, the cost of wastes going outside the plant (enter the environment) is zero. The remaining exergy B_9 will be considered as an external irreversibility of the combustor.

The exergy content of waste flows is internalized to the productive processes that generated it:

$$B_i = \sum_{u \in \mathcal{V}_P} R_{iu}, \quad i \in \mathcal{R} \qquad [5.9]$$

where R_{iu} is the part of waste i generated by process u, and then define its reverse value R_u as the exergy of the wastes produced by process u, as follows:

$$R_u \equiv \sum_{i \in \mathcal{R}} R_{iu}, \quad u \in \mathcal{V}_P \qquad [5.10]$$

Figure 5.8(a) describes the waste internalization of a generic component u; the value R_u is the exergy of process u that becomes a waste stream R_u that is ultimately an external irreversibility I_u^{EX} of the component. According to this idea, the exergy balance equation could be written as:

$$\begin{array}{r} F_u - P_u = I_u^{IN} \\ R_u = I_u^{EX} \\ \hline F_u - (P_u - R_u) = I_u^{IN} + I_u^{EX} \end{array} \qquad [5.11]$$

thus, the term $P_u - R_u$ could be considered as the effective production of the component. For example, in the TGAS plant, the combustor produces gases, whose energy content is 32.99 MW. They are then used in the gas turbine to produce work and in the HRSG to produce steam. At the end of this process, the gases that leave the plant by the stack still have an energy content of 2.17 MW. These gases become external irreversibility $R_1 = I_1^{EX} = 2.17$ of the combustor, and the effective exergy of the gases produced in the combustor is 30.82 MW.

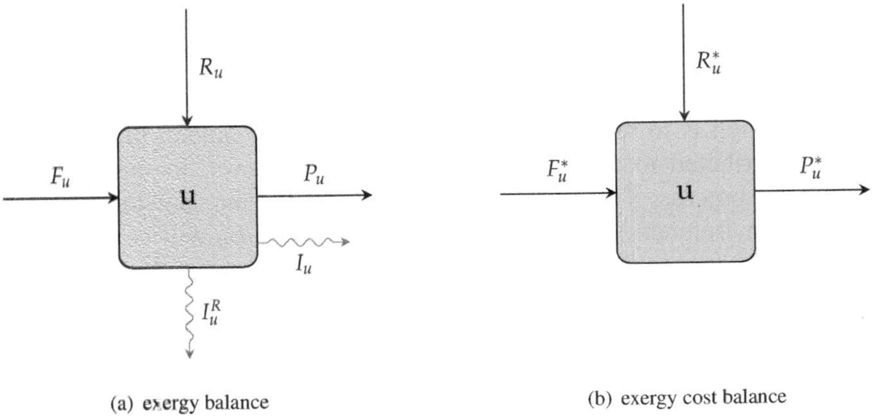

Figure 5.8. *Fuel–product–waste schema for a generic productive component. For a color version of this figure, see www.iste.co.uk/feidt/advances.zip*

On the other hand, we can consider that the cost of the wastes leaving the system is zero because they have no value in the production system, and this exergy cost must be distributed to the component that generates it:

$$B_i^* = \sum_{u \in \mathcal{V}_P} k_i^* \, R_{iu}, \quad i \in \mathcal{R} \qquad [5.12]$$

and the cost of the wastes allocated to a process is given as:

$$R_u^* = \sum_{i \in \mathcal{R}} k_i^* \, R_{iu}, \quad u \in \mathcal{V}_P \qquad [5.13]$$

It means the cost of wastes is internalized and assessed to productive components that generated them. Equation [5.13] is the mathematical formulation of the *waste allocation* rule. Moreover, according to Figure 5.8(b), the *cost conservation* rule must be written as:

$$P_u^* = F_u^* + R_u^*, \quad u \in \mathcal{V}_P \qquad [5.14]$$

Therefore, the production exergy cost of a process P_u^* is equal to the cost of the resources required F_u^* plus the cost of getting rid of wastes, R_u^*.

There is not always a unique criteria to determine the R_{iu} values. In general, the analyst requires a detailed thermodynamic analysis of processes to identify where the wastes are formed. Several papers (Frangopoulos 1987; Gonzalez et al. 2003; Seyyedi et al. 2010; da Silva et al. 2017) analyze the waste flow cost allocation problem from different viewpoints.

Wastes always have a cost, but since they cannot be sold, their cost must be allocated to the final products, or better, to the processes that have produced them. In the same way that there is a cost of formation of products, there is also a cost of formation of waste. In other words, for a fully responsible industry, the company will bear the cost of its waste and endorse this cost on the final product(s). Wastes are obviously a problem for productive systems, but they are also an opportunity for energy saving, by recovering them to be used in other processes (Sciubba 2003; Cecchin et al. 2020).

5.2.4. *Exergy cost computation*

The fuel-product rules offer a procedure for determining the exergy cost of the m flows of an energy plant, building a system of m linear equations with m unknowns, based on the classification of the system streams in fuels, products and wastes. The exergy cost of a flow is an emergent property, that is, we cannot measure it independently of the formation process of the flow. Cost is always linked to the production process. And this process links a set of internal and external flows. Therefore, not only the cost of a flow, but also a complete set of interrelated costs need to be determined. It is interesting to find the costs of final products, but it is also valuable to have them for internal products. Then, the build up of costs for each final product can be traced through the energy system. The same is valid for wastes. They also have a process of cost formation. Their cost is assigned to the final product(s) or is the basis for giving a rational price to convert them into secondary resources in a subsequent process.

EXAMPLE.– Consider the example of the TGAS plant. It has $m = 9$ flows and $n = 5$ processes, one of them is a dissipative unit. Note that air entering the compressor (flow #0) has been removed from the system analysis because its exergy could be considered as zero. To determine the exergy costs of flows of an energy system, it is required to have:

1) the exergy values of the system flows (see Table 5.1);

2) the definition of fuel and product of each process, the efficiency definition (see Table 5.2).

The *cost conservation* rule establishes the cost balance for each process and defined $n = 5$ equations. The *waste* rule establishes that the cost of wastes is formed by adding all resources needed to produce it, and at the end of the analysis, this cost is added to the final cost of the plant production. Therefore, the cost of the flows processed on the dissipative units are assessed to the processes that are generated, by following their irreversibility carrier that identifies their origin, transfer, partial recovering and disposal. In the TGAS plant's case, the cost of the flow dissipated on the stack B_9^* is assessed to the product of the combustion chamber, where gases have been generated.

$$
\begin{aligned}
&B_1\,k_1^* + B_2\,k_2^* + B_9\,k_9^* - B_3\,k_3^* = 0\\
&B_6\,k_6^* - B_2\,k_2^* = 0\\
&B_3\,k_3^* - B_4\,k_4^* - B_6\,k_6^* - B_7\,k_7^* = 0\\
&B_4\,k_4^* - B_5\,k_5^* = 0\\
&B_5\,k_5^* - B_9\,k_9^* = 0
\end{aligned}
$$

The *resources* rule provides one equation per input flow into the plant s_0. In the case of TGAS, $s_0 = 1$, and the cost of natural gas entering combustion chamber is known and is equal to its exergy.

$$k_1^* = 1$$

The F and P rules provide additional equations to allocate costs when a process has more than one output, and generate $s_i - 1$ equations per process. Therefore, the exergy cost rules generate m equations:

$$n + s_0 + \sum_{i=1}^{n} (s_i - 1) = \sum_{i=0}^{n} s_i = m \qquad [5.15]$$

Since the compressor and the combustor only have one output flow, they do not generate additional equations. However, the turbine has three outputs, which entails the definition of two equations. Flow #3 is an unspent fuel of the turbine and according to the *F* rule:

$$k_4^* - k_3^* = 0$$

The other two outputs are co-products, and according to the *P* rule, generate another equation:

$$k_6^* - k_7^* = 0$$

The steam generator (HRSG) also has two outputs, and flow #4 is an unspent fuel, then according to the *F* rule:

$$k_5^* - k_4^* = 0$$

Therefore, we have $m = 9$ equations that allow us to determine the unit cost of flows. The system of linear equations is written in matrix form as:

$$\begin{bmatrix} B_1 & B_2 & -B_3 & 0 & 0 & 0 & 0 & 0 & B_9 \\ 0 & -B_2 & 0 & 0 & 0 & B_6 & 0 & 0 & 0 \\ 0 & 0 & B_3 & -B_4 & 0 & -B_6 & -B_7 & 0 & 0 \\ 0 & 0 & 0 & B_4 & -B_5 & 0 & 0 & -B_8 & 0 \\ 0 & 0 & 0 & 0 & B_5 & 0 & 0 & 0 & -B_9 \\ 1 & 0 & 0 & 0 & 0 & 0 & 0 & 0 & 0 \\ 0 & 0 & 1 & -1 & 0 & 0 & 0 & 0 & 0 \\ 0 & 0 & 0 & 0 & 0 & 1 & -1 & 0 & 0 \\ 0 & 0 & 0 & 1 & -1 & 0 & 0 & 0 & 0 \end{bmatrix} \cdot \begin{bmatrix} k_1^* \\ k_2^* \\ k_3^* \\ k_4^* \\ k_5^* \\ k_6^* \\ k_7^* \\ k_8^* \\ k_9^* \end{bmatrix} = \begin{bmatrix} 0 \\ 0 \\ 0 \\ 0 \\ 0 \\ 1 \\ 0 \\ 0 \\ 0 \end{bmatrix}$$

The system of equations defined according to the FP rules could be written by hand or using a matrix approach algorithm described in some previous studies (Torres et al. 2007; Querol et al. 2013; Lazzaretto and Tsatsaronis 2006; Valero et al. 1987).

The solution of the above system of equations provides the unit cost of the flows (see Table 5.4). More examples of thermoeconomic analysis applications to energy conversion systems can be found in Kumar (2017); Keshavarzian et al. (2017).

5.3. Structural theory of thermoeconomics

Over the years, the clarity and generality of the theory of exergy cost have allowed it to be widely used by practitioners. Moreover, important applications have been developed for the real-time diagnosis of industrial plants, but questions have also been raised that could not be solved within the same theory. In the following, we will explain the advantages and disadvantages of the theory and the need for a more general theory.

First, based on its definition, the exergy cost depends on the plant disaggregation scheme. In other words, it cannot render a precise value because it depends on the analyst choice. Let us suppose a chemical plant that produces ethylene and electricity in different processes but in the same plant. Obviously, in this aggregated scheme, the P or F rules would not apply. We have to analyze the formation of the products, the type of streams, as well as where the exergy is consumed. This detailed analysis provides a productive structure to which the rules can be applied in practical terms. This detailed definition is more precise; however, there is a risk of not having measured data to corroborate the sensitivity of the costs obtained to variations in these data. There may be large discrepancies among practitioners with regard to the definition of production structures and the method of disaggregation of processes.

Second, the definition of exergy efficiency also depends on the analyst. For example, in a countercurrent heat exchanger, a stream becomes cooled, and another becomes heated. Hence, it heats, cools, or both, then, which stream assumes the production cost? Moreover, not all components are intended to save exergy (productive components). Some other components of a plant belong to the control system, such as throttle valves, or serve to mitigate (the exergy of) waste. Therefore, it must be determined how to allocate costs to the components that process the waste.

Third, the concept of exergy cost only measures how the physical resources consumed by the production plant are distributed. By avoiding monetary units to measure costs, external fluctuations of costs not attributable to the plant's behavior are eliminated. This allows us to focus on the diagnosis of malfunctions that occur in industrial processes (Valero et al. 2004). Such an idea opened up a new scientific-technological field because no energy system deteriorates from a single cause, but from several at once, with different intensities and chains of causality that induce new malfunctions. Thermoeconomic diagnosis using exergy costs can be implemented online and is very useful when analyzing complex plants. The disaggregation of any industrial energy system into sub-systems allows the research of the formation of physical costs, predicts failures and helps maintain the plant in optimal operating conditions. Exergy costs are sensible to detect failures but still depend on the plant's reference state, which can be taken as fixed, or variable depending on external environmental conditions. In fact, the exergy and its related cost depend on the chosen reference environment. So the results will never be absolute (Valero and Torres 2020).

Moreover, are the costs obtained average costs or marginal costs? The exergy of a product measures its potential work with respect to a given environment, and the exergy cost measures how many irreversibilities have occurred in the manufacture of that product plus its remaining exergy.

The exergy cost can be considered an *exergy backpack* or *exergy footprint* of a stream or product, that is, the past exergy synopsis (history) of the stream, whereas its exergy is the future energy potential of such a stream or its remaining capacity to produce work. Finally, can the exergy costs predict future degradations or describe only past irreversibilities?

Despite these objections, the analyst aims to obtain reasonable costs. Thus, it is necessary to disaggregate the system as far as common sense advises. However, if costs depend on the disaggregation scheme, what scheme is reasonable? Second, the definition of efficiency for each component requires caution. Is it possible to find a more rational definition of efficiency? Third, choosing exergy is apparently better than other exergy-like functions because it provides a proxy for the locally usable energy. Are there better functions that predict the behavior of components? In fact, costs should be as useful as possible in predicting how many additional resources are needed to compensate for small perturbations of a system. The structural theory is a step further in solving such questions.

Any energy system that consumes resources and manufactures products can be characterized by a thermodynamic function G of the generic type, $G = m(h - T_x\, s)$, where m is the mass, h is the enthalpy, s is the entropy with respect to a certain reference and T_x is the temperature to choose under certain conditions. When T_x is zero, the function G is the enthalpy; when T_x is the ambient temperature, G is the exergy; if T_x is the temperature of the system, G is the Gibbs free energy function.

The behavior of any u component can be described by functions of the type:

$$G_i = g_i\left(\{x\}, \{G_j\}\right) \qquad i \in \mathcal{E}_u \qquad [5.16]$$

where $i \in \mathcal{E}_u$ are the input flows of component u and $j \in \mathcal{S}_u$ the output flows of component u, and $\{x\}$ is the set of its internal governing parameters, for instance, internal efficiencies, pressure relationships, temperature increases, geometric, structural, fluid dynamic or heat transfer parameters and so on. If the system environment is the component "0", then $\mathcal{E}_0$ is the set of the system outputs and $\mathcal{S}_0$ is the set of input resources to the system.

Each flow G_i outputs either from the environment or from a system component and inputs to another component or the environment. The final products of the system are the inputs to the environment. Therefore, in this case, the characteristic function is defined as:

$$G_i = \omega_i \qquad i \in \mathcal{E}_0 \tag{5.17}$$

For example, in Table 5.3, the general characteristic equations of the TGAS plant are shown according to its physical structure.

Nr	Component	Entry	Outlet	Equation
1	COMB	[1,2]	[3]	$G_1 = g_1(x, G_2)$ $G_2 = g_2(x, G_2)$
2	COMP	[6]	[2]	$G_6 = g_6(x, G_2)$
3	GTUR	[3]	[4,6,7]	$G_3 = g_3(x, G_4, G_6, G_7)$
4	HRSG	[3]	[4,8]	$G_4 = g_4(x, G_5, G_8)$
5	STCK	[5]	[9]	$G_5 = g_5(x, G_9)$
0	ENV	[7, 8, 9]	[1]	$G_7 = \omega_7$ $G_8 = \omega_8$ $G_9 = \omega_9$

Table 5.3. *Characteristic equations of TGAS plant*

The functions characterising the inlet and outlet flows are extensive magnitudes, typically a mass flow rate multiplied by a specific energy-like function describing the flow quality. The internal variables appearing in the thermoeconomic model depend on the behavior of the subsystem, and they are presumably independent of mass flow rates. It implies that relations like efficiencies or pressure and temperature ratios which are mainly independent of the quantity of the exiting flows can be used as internal parameters. These functions could be represented using a graph, like the one shown in Figure 5.9 for a TGAS plant.

Depending on the analysis scope, a subsystem can be identified as a separate piece of equipment, a part of a device, several process units, or even the whole plant. Sometimes the objective consists of analyzing a plant in great detail. In this case, it is practically advisable to identify each subsystem with a separate physical process (heat transfer, pressure increase or decrease, and chemical mixture or reaction) in order to locate and quantify, separately if possible, each thermal, mechanical, and chemical irreversible process occurring in the plant. Thus, thermoeconomics always performs a systemic analysis, no matter how complex the system, oriented toward locating and quantifying the energy conversion efficiency and the process of energy quality degradation. It is not within the scope of thermoeconomics to model the process units' behavior, which is done by the mathematical equations of the physical model.

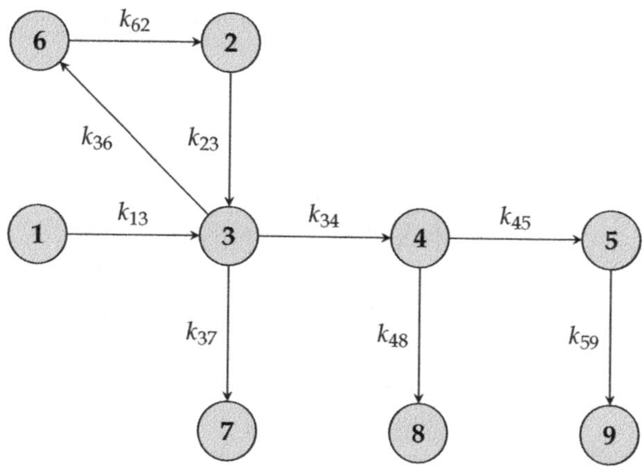

Figure 5.9. *Graph representation of TGAS characteristic equations. For a color version of this figure, see www.iste.co.uk/feidt/advances.zip*

Even though it is not the objective of thermoeconomics to simulate the subsystems' behavior, it is essential to build a thermoeconomic model with a physical meaning. This is the reason, as already explained, for defining different thermoeconomic models for the same plant. Depending on the aggregation level and the nature of thermoeconomic equations, the model will contain physical information about the actual system behavior with different degrees of detail. The results obtained from a very rough thermoeconomic model that is not sensitive to any physical detail related to the plant's actual behavior will probably be useless.

5.3.1. *Marginal cost equations*

The cost of system resources G_0 can then be expressed, remembering the resources rule of exergy cost, by the following equation:

$$G_0 = \sum_{i \in \mathcal{S}_0} G_i \qquad [5.18]$$

Once the characteristic equations of the system have been established, the marginal cost of the system flows is defined by:

$$k_i^* \equiv \left(\frac{\partial G_0}{\partial G_i}\right) \qquad [5.19]$$

and it can be calculated using the chain rule of derivation as:

$$\begin{aligned} &\left(\frac{\partial G_0}{\partial G_j}\right) = 1 && j \in \mathcal{S}_0 \\ &\left(\frac{\partial G_0}{\partial G_j}\right) = \sum_{i \in \mathcal{E}_u} \left(\frac{\partial G_0}{\partial G_i}\right)\left(\frac{\partial g_i}{\partial G_j}\right) && j \in \mathcal{S}_u \end{aligned} \qquad [5.20]$$

Equation [5.20] evaluates the additional consumption of system resources when an additional unit of flow G_i is produced, under the condition that internal variables x do not vary throughout the process.

If we denote the marginal consumption k_{ij} as the amount of local resource G_i required to produce an additional unit of the flow G_j:

$$k_{ij} \equiv \left(\frac{\partial g_i}{\partial G_j}\right) \qquad [5.21]$$

In the case that G is exergy, k_{ij} will become the marginal consumption of flow G_i.

Then, using this nomenclature, [5.20] can be rewritten as:

$$k_i^* = \begin{cases} \sum_{j \in \mathcal{E}_u} k_j^* \, k_{ji}, & i \in \mathcal{S}_u, \quad u \in \mathcal{N} \\ 1, & i \in \mathcal{S}_0 \end{cases} \qquad [5.22]$$

These equations can be written in matrix notation as follows:

$$(\mathbf{U}_m - {}^{\mathrm{t}}\langle\mathbf{G}\rangle)\,\mathbf{k}^* = \mathbf{c}_0 \qquad [5.23]$$

where $\langle\mathbf{G}\rangle$ is a $(m \times m)$ matrix containing the Jacobian of the characteristic equations, its elements are the marginal exergy consumptions k_{ij}, $\mathbf{k}^*$ is a vector $(m \times 1)$ which contains the marginal costs of each flow, and $\mathbf{c}_0$ is a $(m \times 1)$ vector related to the cost of the system resource, defined as:

$$c_{0,i} = \begin{cases} 1 & i \in \mathcal{S}_0 \\ 0 & \text{otherwise} \end{cases} \qquad [5.24]$$

Given the boundaries of the system, the marginal cost of the resources entering the plant is known and equals 1, because there is no energy quality degradation, the amount of resources consumed to obtain each system resource is its own energy content. Note that this result is equivalent to the *resources rule* of the ECT.

Once the characteristic equations and then the marginal consumption for each component are known, the marginal costs k_i^* can be obtained by solving the system of linear equations [5.23]. This set of equations, also represented by the characteristic diagram (see Figure 5.9), shows the processes of cost formation, that is, how the flow costs are generated through the system components.

As k^* was defined as a marginal cost when production is modified, we can also obtain the marginal cost when the internal variables x are modified. Applying the chain rule, we get:

$$\left(\frac{\partial G_0}{\partial x}\right) = \sum_{u\in\mathcal{V}}\sum_{i\in\mathcal{E}_u}\left(\frac{\partial G_0}{\partial G_i}\right)\left(\frac{\partial g_i}{\partial x}\right) = \sum_{u\in\mathcal{V}}\sum_{i\in\mathcal{E}_u} k_i^*\left(\frac{\partial g_i}{\partial x}\right) \qquad [5.25]$$

This equation is the generalization *fuel impact* formula presented in the previous section and expresses the effect on additional resource consumption when an internal parameter x is modified. It shows how the marginal cost of flows k_i^* is used to weigh a local malfunction given by $\partial g_i/\partial x$, and it is the basis for the thermoeconomic diagnosis (Valero et al. 2004) and optimization (Valero et al. 1994).

5.3.2. *Linear model of characteristic equations*

The general model of marginal cost explained above provides a formal and general way of cost computation and leads to important remarks that should be analyzed. What is the relationship between marginal costs, used for example, in the optimization of energy systems (Frangopoulos 1987; von Spakovsky and Evans 1988), and average costs provided by cost accounting methods like the ECT?

The characteristic functions (equation [5.16]) generally do not relate inputs and outputs linearly. Because the only degree of freedom is the choice of T_x (i.e. the choice of energy, exergy, Gibbs function, or any other function), it is likely that the complexity of the equation cannot be reduced to a linear relationship. However, such a continuous function allows for a Taylor series approximation. Besides, by choosing sufficiently small linear intervals, second-order terms can be discarded. The smaller the interval, the closer the proposed behavior is to reality.

Let us consider the characteristic equations as linear functions, with respect to the magnitude that represents the flow streams:

$$g_i(\mathbf{G}, \mathbf{x}) = \begin{cases} \sum\limits_{j \in \mathcal{S}_u} \gamma_{ij}(x)\, G_j & i \in \mathcal{E}_u \\ \omega_i \; (const.) & i \in \mathcal{E}_0 \end{cases} \qquad [5.26]$$

The outflows magnitudes here are independent variables with respect to the internal parameters. If this condition is achieved, then the marginal consumption becomes like the coefficients of the characteristic equations, that is:

$$\gamma_{ij} = \left(\frac{\partial g_i}{\partial G_j}\right) = k_{ij} \qquad [5.27]$$

Therefore, [5.26] can be written in matrix form as:

$$(\mathbf{U}_m - \langle \mathbf{G} \rangle)\, \mathbf{G} = \mathbf{G}_S \qquad [5.28]$$

where $\mathbf{G}_s$ is a $(m \times 1)$ the vector of external products, defined as:

$$G_{S,i} = \begin{cases} \omega_i & i \in \mathcal{E}_0 \\ 0 & \text{otherwise} \end{cases} \qquad [5.29]$$

Note that matrix $\langle \mathbf{G} \rangle$ is obtained once the linear characteristic equations are defined. This fact allows the matrix $\langle \mathbf{G} \rangle$ to be obtained, and then the marginal costs, once the linear characteristic equations have been defined. In linear algebra, these structures are called the primal and dual problem, defined in our case as:

$$[\text{Primal}] \equiv \begin{cases} G_0 = {}^t\mathbf{c}_0 \cdot \mathbf{G} \\ \mathbf{G} = \langle \mathbf{G} \rangle \, \mathbf{G} + \mathbf{G}_s \end{cases} \qquad [\text{Dual}] \equiv \begin{cases} G_0 = {}^t\mathbf{k}^* \cdot \mathbf{G}_s \\ {}^t\mathbf{k}^* = {}^t\langle \mathbf{G} \rangle \, {}^t\mathbf{k}^* + \mathbf{c}_0 \end{cases} \qquad [5.30]$$

EXAMPLE.– The system of equations, [5.23], to determine the marginal cost of the flows of a TGAS plant is:

$$\begin{bmatrix} 1 & 0 & 0 & 0 & 0 & 0 & 0 & 0 & 0 \\ 0 & 1 & 0 & 0 & 0 & -k_{62} & 0 & 0 & 0 \\ -k_{12} & -k_{23} & 1 & 0 & 0 & 0 & 0 & 0 & 0 \\ 0 & 0 & -k_{34} & 1 & 0 & 0 & 0 & 0 & 0 \\ 0 & 0 & 0 & -k_{45} & 1 & 0 & 0 & 0 & 0 \\ 0 & 0 & -k_{36} & 0 & 0 & 1 & 0 & 0 & 0 \\ 0 & 0 & -k_{37} & 0 & 0 & 0 & 1 & 0 & 0 \\ 0 & 0 & 0 & -k_{48} & 0 & 0 & 0 & 1 & 0 \\ 0 & 0 & 0 & 0 & -k_{59} & 0 & 0 & 0 & 1 \end{bmatrix} \cdot \begin{bmatrix} k_1^* \\ k_2^* \\ k_3^* \\ k_4^* \\ k_5^* \\ k_6^* \\ k_7^* \\ k_8^* \\ k_9^* \end{bmatrix} = \begin{bmatrix} 1 \\ 0 \\ 0 \\ 0 \\ 0 \\ 0 \\ 0 \\ 0 \\ 0 \end{bmatrix}$$

The matrix $\langle \mathbf{G} \rangle$ is the adjacency matrix of the characteristic equation graph, as shown in Figure 5.9. This graph allows us to analyze the flows' cost formation processes, following the chain rule of derivation.

5.4. Structural theory and exergy cost

In this section, we will prove that the ECT is a particular case of the ST when the characteristic equations are derived from the definition of processes efficiency of the system.

First, we will prove that when using the characteristic equations concerning the extensive magnitude G, the system of equations required for the calculation of marginal and average costs coincide. This is a very important result because the marginal and average costs can be calculated using the same procedure.

To illustrate this fact, let us consider a generic component or subsystem of a plant with several inlet and outlet flows. For the sake of simplicity, we shall consider a general subsystem with two inlet flows and two outlet flows, as depicted in Figure 5.10

The characteristic equations that describe the behavior of this component are:

$$G_i = \sum_{j \in S_u} \kappa_{ij} G_j \quad i \in \mathcal{E}_u \qquad [5.31]$$

According to Figure 5.10, the input resources can be decomposed as:

$$G_1 = k_{13}G_3 + k_{14}G_4$$
$$G_2 = k_{23}G_3 + k_{24}G_4$$

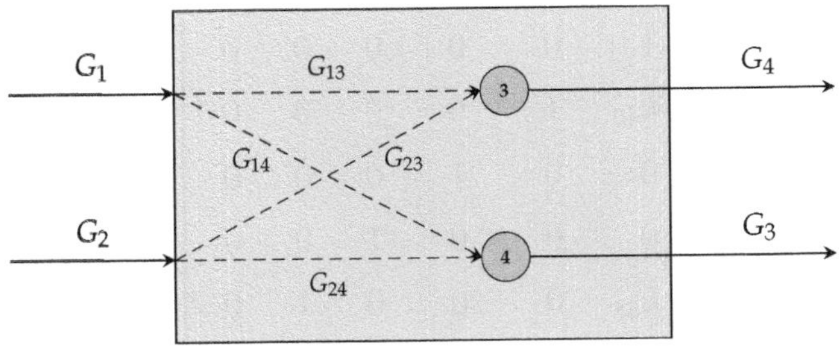

Figure 5.10. *General component scheme*

These equations relate the amount of inlet resources (G_1, G_2) consumed to obtain each one of the outlet flows (G_3, G_4). Correspondingly, the total amount of resources consumed for the whole system for obtaining G_3 and G_4 is:

$$\bar{k}_1^* \, G_1 = \bar{k}_1^* \, k_{13} \, G_3 + \bar{k}_1^* \, k_{14} \, G_4$$
$$\bar{k}_2^* \, G_2 = \bar{k}_2^* \, k_{23} \, G_3 + \bar{k}_2^* \, k_{24} \, G_4$$

where $\bar{k}_i^*$ now represents the average unit cost of the flow. If the component is considered to be made up of two subsystems, one per output, then the equation modeling for each subsystem is:

$$G_{ij} = k_{ij} \, G_j \qquad i \in \mathcal{E}_u, j \in S_u \qquad [5.32]$$

These equations represent the resources needed to produce the outputs. Therefore, the total amount of resources consumed to obtain G_3 and G_4 are:

$$\begin{aligned}\bar{k}_1^*\, G_{13} + \bar{k}_2^*\, G_{23} &= \bar{k}_1^*\, \kappa_{13}\, G_3 + \bar{k}_2^*\, \kappa_{23}\, G_3 \\ \bar{k}_1^*\, G_{14} + \bar{k}_2^*\, G_{24} &= \bar{k}_1^*\, \kappa_{14}\, G_4 + \bar{k}_2^*\, \kappa_{24}\, G_4\end{aligned} \qquad [5.33]$$

Hence, the average cost of the outlet flows is:

$$\begin{aligned}\bar{k}_3^* &= \frac{\bar{k}_1^*\kappa_{13}G_3 + \bar{k}_2^*\kappa_{23}G_3}{G_3} = \bar{k}_1^*\kappa_{13} + \bar{k}_2^*\kappa_{23} \\ \bar{k}_4^* &= \frac{\bar{k}_1^*\kappa_{14}G_4 + \bar{k}_2^*\kappa_{24}G_4}{G_4} = \bar{k}_1^*\kappa_{14} + \bar{k}_2^*\kappa_{24}\end{aligned} \qquad [5.34]$$

or in general:

$$\bar{k}_j^* = \sum_{i \in \mathcal{E}_0} \bar{k}_i^*\, \kappa_{ij} \qquad [5.35]$$

According to [5.22], the average costs of the output of a component coincide with their marginal costs. Both kinds of costs are the same because the equations modeling the component are linear functions with respect to the extensive magnitudes that characterize the outlet flows.

Applying [5.22] and [5.26], the cost balance for each system component is satisfied:

$$\begin{aligned}\sum_{i \in \mathcal{E}_u} k_i^*\, G_i = \sum_{i \in \mathcal{S}_u} k_i^* \left(\sum_{j \in \mathcal{E}_u} k_{ij}\, G_j \right) &= \sum_{j \in \mathcal{S}_u} \left(\sum_{i \in \mathcal{E}_u} k_i^*\, k_{ij} \right) G_j \\ &= \sum_{j \in \mathcal{S}_u} k_j^*\, G_j\end{aligned} \qquad [5.36]$$

Equation [5.36] can be expressed in a matrix form as:

$$\mathbf{A} \cdot \mathbf{B}^* = \mathbf{0} \qquad [5.37]$$

where $\mathbf{A}$ is the incidence matrix of the system, defined as:

$$a_{ij} = \begin{cases} 1 & \text{if} \quad j \in \mathcal{E}_i \\ -1 & \text{if} \quad j \in \mathcal{S}_i \\ 0 & \text{otherwise} \end{cases} \qquad [5.38]$$

For example in case of TGAS plant, the incidence matrix is:

$$\mathbf{A} = \begin{bmatrix} 1 & 1 & -1 & 0 & 0 & 0 & 0 & 0 & 0 \\ 0 & -1 & 0 & 0 & 0 & 1 & 0 & 0 & 0 \\ 0 & 0 & 1 & -1 & 0 & -1 & -1 & 0 & 0 \\ 0 & 0 & 0 & 1 & -1 & 0 & 0 & -1 & 0 \\ 0 & 0 & 0 & 0 & 1 & 0 & 0 & 0 & -1 \end{bmatrix}$$

In other words, [5.37] explains the *cost conservation* rule of ECT. This result is quite remarkable because the following equations summarize the equilibrium thermodynamics straightforwardly for open systems:

$$\mathbf{A} \cdot \mathbf{M} = \mathbf{0} \qquad \text{Mass Balance} \qquad [5.39]$$

$$\mathbf{A} \cdot \mathbf{H} = \mathbf{0} \qquad \text{Energy Balance} \qquad [5.40]$$

$$\mathbf{A} \cdot \mathbf{B} = \mathbf{I} \qquad \text{Exergy Balance} \qquad [5.41]$$

$$\mathbf{A} \cdot \mathbf{B}^* = \mathbf{0} \qquad \text{Cost Balance} \qquad [5.42]$$

where **M**, **H** and **B** are the mass, enthalpy and exergy vectors of the material streams: m_i, $m_i\, h_i$ and $m_i(h_i - T_0\, s_i)$; if there are the heat flows: *zero*, Q and $Q(1 - T_0/T)$; and *zero*, W , W if there are working flows, respectively. Note the parallelism with respect to the balance cost equation [5.37].

The application of equations [5.22] and [5.26] explains the previously mentioned rules F and P as a particular case. For example, for a turbine (see Figure 5.3), the following defined efficiency, is considered:

$$B_1 = B_2 + k\, B_3$$

Suppose now, the characteristic equation of the turbine coincides with the exergy definition of the turbine. In the simulations of the turbine for a reasonably small working interval, k can be considered constant. By applying the unit cost equation [5.22], the costs of outputs #2 and #3 are obtained:

$$k_3^* = k\, k_1^* \qquad \text{and} \qquad k_2^* = k_1^*$$

hence, the result is the *recoverable fuel rule*, that is, if a resource has not been consumed (stream 2), its unit cost is equal to that of the input resource (stream 1).

However, if a typical electric power distribution system is considered (see Figure 5.4), its efficiency can be expressed as follows:

$$B_1 = k\,B_2 + k\,B_3$$

Again, it is assumed that the characteristic equation of the distribution system coincides with its exergy definition. If k is constant, applying the unit cost equation [5.26] results in:

$$k_2^* = k\,k_1^* \qquad \text{and} \qquad k_3^* = k\,k_1^* \qquad [5.43]$$

which explains why the *co-production rule* applies, that is, the unit costs of products of the same nature produced in the same process are equal.

In the case of the *waste rule,* equations [5.9] and [5.13] are primal-dual, and the corresponding characteristic equation is:

$$B_i = \sum_{\substack{u \in \mathcal{V}_P \\ j \in \mathcal{S}_u}} R_{ij} = \sum_{\substack{u \in \mathcal{V}_P \\ j \in \mathcal{S}_u}} \rho_{ij}\,B_j, \quad i \in \mathcal{R} \qquad [5.44]$$

where ρ_{ij} is the amount of wasted exergy B_i per exergy unit of output stream B_j of the productive component u.

In the case of the TGAS example, the characteristic equation is:

$$B_9 = \rho_{93}\,B_3,$$

where $\rho_{93} \equiv B_9/B_3$ is the ratio of unspent combustion gases per total combustion gases.

5.4.1. *The flow-process model*

In general, given the definition of efficiency for each of the system processes, it is possible to construct their corresponding characteristic equations and calculate the exergy costs of the system flows.

Each component of the system has one or more fuel and product streams; let $\mathcal{F}_u = \{f_1, \ldots, f_{l_u}\}$ and $\mathcal{P}_u = \{p_1, \ldots, p_{h_u}\}$ to be the sets of fuel and product streams of the process u, where l_u is the number of fuel streams, and h_u is the number of product streams of process u. Finally, let

$\mathcal{L} = \{\ell_1, \ldots, \ell_{nr}\}$ be the set of all the productive (fuel and products) groups of the system.

On the other hand, each stream $\ell \in \mathcal{L}$ could have several input and output flows. Let us denote $\mathcal{E}_l$ and $\mathcal{S}_l$ as the sets of input and output streams of the productive group ℓ. A fuel stream must have at least one input flow, and a product stream must have at least one output flow.

Defining system boundaries is essential to determine the system costs. For this purpose, the environment is defined as an additional process, complementary to the global plant. The system outputs are the fuel of the environment $\mathcal{F}_0$, and the system resources are the product of the environment $\mathcal{P}_0$. A productive group must be defined for each final product and each external resource to ensure the information about them.

Each productive group $\ell \in \mathcal{L}$ has associated a measure corresponding to the exergy processed. Let us denote $F_{u,l}$ as the exergy of the fuel stream ℓ of the process u, therefore:

$$F_{u,l} = \sum_{i \in \mathcal{E}_l} B_i - \sum_{j \in \mathcal{S}_l} B_j, \geq 0 \qquad \ell \in \mathcal{F}_u \qquad [5.45]$$

Similarly, in case of a product stream, let us denote $P_{u,l}$ as the exergy product stream ℓ of the process u.

$$P_{u,l} = \sum_{j \in \mathcal{S}_l} B_i - \sum_{i \in \mathcal{E}_l} B_i, \geq 0 \qquad \ell \in \mathcal{P}_u \qquad [5.46]$$

Finally, the fuel F_u and product P_u of each process u is defined as the sum of their corresponding streams, and by [5.2]: $F_u \geq P_u$.

$$F_u = \sum_{l \in \mathcal{F}_u} F_{u,l} > 0 \quad \text{and} \quad P_u = \sum_{l \in \mathcal{P}_u} P_{u,l} > 0 \qquad [5.47]$$

The productive structure of a system is based on the underlying idea that *the processes use flows to produce flows*. The flow-process approach has two sides:

On the use side, the inflows of a fuel stream are used as resources of the process, and additionally, part of the exergy of the stream, not used by the process, becomes the output flow of the productive group. Given a fuel stream ℓ of the process u, from [5.45]:

$$\sum_{i \in \mathcal{E}_l} B_i = F_{u,l} + \sum_{j \in \mathcal{S}_l} B_j \tag{5.48}$$

Let $FE_{u,l}$ be the input exergy of a fuel stream and $PS_{u,l}$ be the output exergy of a product stream. We define the *input exergy ratio* x_i as:

$$x_i = \begin{cases} \dfrac{B_i}{FE_{u,l}} & i \in \mathcal{E}_l,\ \ell \in \mathcal{F}_u \\ \\ \dfrac{B_i}{PS_{u,l}} & i \in \mathcal{E}_l,\ \ell \in \mathcal{P}_u \end{cases} \tag{5.49}$$

therefore:

$$B_i = x_i\, F_{u,l} + x_i \sum_{j \in \mathcal{S}_l} B_j, \quad i \in \mathcal{E}_l\ l \in \mathcal{F}_u \tag{5.50}$$

On the production side, the outflows of a stream come, partly from the product of the process and optionally by flows that come from other processes, from [5.46], and this results in:

$$\sum_{j \in \mathcal{S}_l} B_j = P_{u,l} + \sum_{i \in \mathcal{E}_l} B_i \tag{5.51}$$

Let y_j be the *bifurcation ratio*, defined as:

$$y_j = \begin{cases} \dfrac{B_j}{PS_{u,l}} & j \in \mathcal{S}_l,\ \ell \in \mathcal{P}_u \\ \\ \dfrac{B_j}{FE_{u,l}} & j \in \mathcal{S}_l,\ \ell \in \mathcal{F}_u \end{cases} \tag{5.52}$$

therefore:

$$B_j = y_j\, P_{u,l} + y_j \sum_{i \in \mathcal{E}_l} B_i, \quad j \in \mathcal{S}_l,\ l \in \mathcal{P}_u \tag{5.53}$$

Let $\mathbf{F}$ and $\mathbf{P}$ be $(n \times 1)$ vectors, whose elements are the fuel and product of the processes, and $\mathbf{B}$ a $(m \times 1)$ vector, whose elements are the exergy of the flows. We will denote by ${}^{t}\mathbf{u}_n \equiv (1, \ldots, 1)$ the unitary row vector of dimension n, therefore, the previous relationships can be written in matrix form as follows:

$$\begin{aligned} \mathbf{F} &= {}^{t}[\mathbf{F}]\,\mathbf{u}_m \\ \mathbf{P} &= [\mathbf{P}]\,\mathbf{u}_m \\ \mathbf{B} &= \boldsymbol{\omega}_0 + [\mathbf{F}]\,\mathbf{u}_n + [\mathbf{V}]\,\mathbf{u}_m = \boldsymbol{\nu}_0 + {}^{t}[\mathbf{P}]\,\mathbf{u}_n + {}^{t}[\mathbf{V}]\,\mathbf{u}_m \end{aligned} \qquad [5.54]$$

where $[\mathbf{F}]$ is a $(m \times n)$ matrix, whose elements, $f_{iu} = x_i\, F_{u,l}$, $i \in \mathcal{E}_l$, $\ell \in \mathcal{F}_u$, represent the exergy of the input flow i used as fuel in process u.

Similarly, let $[\mathbf{P}]$ be a $(n \times m)$ matrix , whose elements, $p_{uj} = y_j\, P_{u,l}$, $j \in \mathcal{S}_l$, $\ell \in \mathcal{P}_u$, represent the exergy value of the output flow j used as product of process u.

Finally, let $[\mathbf{V}]$ be a $(m \times m)$, whose elements $v_{ij} = x_i\, B_j = y_j\, B_i$ represent the exergy of the input flow $i \in \mathcal{E}_l$, which becomes the output flow $j \in \mathcal{S}_l$. These elements are defined only if the stream has both input and output flows.

The table in Figure 5.11(a), called a flow-process table, summarizes the information provided by equation [5.54].

The vector $\boldsymbol{\nu}_0$ $(m \times 1)$ contains the exergy values of the external resources, and the vector $\boldsymbol{\omega}_0$ $(m \times 1)$ contains the exergy values of the system outputs:

$$\nu_{0,i} = \begin{cases} B_i & i \in \mathcal{S}_0 \\ 0 & \text{otherwise} \end{cases} \qquad \omega_{0,i} = \begin{cases} B_i & i \in \mathcal{E}_0 \\ 0 & \text{otherwise} \end{cases} \qquad [5.55]$$

The rows in the table in Figure 5.11(a) reflect the use of inputs flows in processes, meanwhile, the columns show the output flows produced by the process.

Waste flows are identified as all the unwanted and unrecoverable outputs of a system. Let us denote $\mathcal{R}$ as the subset of output system flows $\mathcal{E}_0$ which are wastes of the system. Therefore, the system outputs can be split into two parts:

$$\boldsymbol{\omega}_0 = \boldsymbol{\omega}_s + \boldsymbol{\omega}_r \qquad [5.56]$$

outputs (inputs)

	$\mathbf{0}$	$\boldsymbol{\nu}_0$	F_T
$\mathbf{0}$	$\mathbf{0}$	[**P**]	**P**
$\boldsymbol{\omega}_0$	[**F**]	[**V**]	**B**
P_T	**F**	**B**	

(a) Flow-Process Table

outputs (inputs)

	$\mathbf{0}$	$\mathbf{0}$	$\mathbf{0}$
$\mathbf{0}$	$\mathbf{0}$	$\mathbf{0}$	$\mathbf{0}$
$\mathbf{0}$	[**R**]	$\mathbf{0}$	$\boldsymbol{\omega}_r$
$\mathbf{0}$	**R**	$\mathbf{0}$	

(b) Waste-Process Table

Figure 5.11. *Flows-Processes and Waste-Process Tables*

Let $\boldsymbol{\omega}_r$ be a $m \times 1$ vector containing the exergy of waste flows and $\boldsymbol{\omega}_s$ a $m \times 1$ vector containing the exergy of the final products:

$$\omega_{r,i} = \begin{cases} E_i & i \in \mathcal{E}_0 \cup \mathcal{R} \\ 0 & \text{otherwise} \end{cases} \qquad \omega_{s,i} = \begin{cases} E_i & i \in \mathcal{E}_0 \cup \bar{\mathcal{R}} \\ 0 & \text{otherwise} \end{cases} \qquad [5.57]$$

Let [**R**] be a $(n \times n)$ matrix, called *waste-process table*, whose elements R_{iu} represent the exergy of waste flow $i \in \mathcal{R}$ generated by the productive process $u \in \mathcal{V}_P$. This matrix verifies the following properties:

$$\boldsymbol{\omega}_r = [\mathbf{R}]\, \mathbf{u}_n \qquad [5.58]$$

$$\mathbf{R} = {}^{\mathsf{t}}[\mathbf{R}]\, \mathbf{u}_m \qquad [5.59]$$

where **R** is a $(1 \times n)$ vector whose elements contain the exergy of the wastes generated by each process. The waste-process table in Figure 5.11(b), describes the cost formation process of wastes, integrated as an additional layer of the productive structure. A more detailed explanation of the *flow-process model* and applications can be found in Torres and Valero (2021).

In the case of TGAS, there is only a waste flow: B_9. Its exergy has two components: thermal, because the temperature of the gases is higher than the environment temperature, and chemical because the composition of gases is different from the reference environment. The combustion process is the origin of the thermal and chemical exergy of gases. Therefore, the cost of gases must be assessed to the combustor. Then matrix [**R**] has only one entry $R_{9,1} = B_9$.

EXAMPLE.– The flow-process matrices of the TGAS plant, represented in sparse format (row, col, value), are:

$$[\mathbf{F}] + [\mathbf{R}]\ (9 \times 5) = \begin{bmatrix} 1 & 3 & 4 & 5 & 6 & 9 \\ 1 & 3 & 4 & 5 & 2 & 1 \\ B_1 & B_3 - B_4 & B_4 - B_5 & B_5 & B_6 & B_9 \end{bmatrix}$$

$$[\mathbf{P}]\ (5 \times 9) = \begin{bmatrix} 1 & 2 & 3 & 3 & 4 & 5 \\ 3 & 2 & 6 & 7 & 8 & 9 \\ B_3 - B_2 & B_2 & B_6 & B_7 & B_8 & B_9 \end{bmatrix}$$

$$[\mathbf{V}]\ (9 \times 9) = \begin{bmatrix} 2 & 3 & 4 \\ 3 & 4 & 5 \\ B_2 & B_4 & B_5 \end{bmatrix}$$

Observe that the lower left part of the table $[\mathbf{F}]$ contains the fuel definition of the processes, and the upper right part $[\mathbf{P}]$ contains the product definition. The lower right part $[\mathbf{V}]$ is used to close the balance of the input–output flows.

5.4.2. *The characteristic equations of exergy cost theory*

The characteristic matrix, associated with the ECT, can be obtained from the flow-process table relationships. Consider the column scaled matrices:

$$\gamma_F(m \times n) \equiv [\mathbf{F}]\,\hat{\mathbf{P}}^{-1} \qquad \gamma_R(m \times n) \equiv [\mathbf{R}]\,\hat{\mathbf{P}}^{-1}$$
$$\gamma_P(n \times m) \equiv [\mathbf{P}]\,\hat{\mathbf{B}}^{-1} \qquad \gamma_V(m \times m) \equiv [\mathbf{V}]\,\hat{\mathbf{B}}^{-1}$$

where $\hat{\mathbf{B}}$ and $\hat{\mathbf{P}}$ are diagonal matrices from vectors $\mathbf{B}$ and $\mathbf{P}$, respectively. By applying them to [5.54], we obtain:

$$\begin{aligned} \mathbf{P} &= \gamma_P\,\mathbf{B} \\ \mathbf{B} &= \omega_0 + \gamma_F\,\mathbf{P} + \gamma_V\,\mathbf{B} \end{aligned} \qquad [5.60]$$

By equation [5.56], the system output flows can be split into final products and wastes:

$$\omega_0 = \omega_s + \gamma_R\gamma_P\,\mathbf{B} \qquad [5.61]$$

therefore:

$$\mathbf{B} = \boldsymbol{\omega}_s + (\gamma_F + \gamma_R)\,\gamma_P\,\mathbf{B} + \gamma_V\,\mathbf{B} \qquad [5.62]$$

If the characteristic matrix is taken as:

$$\langle \mathbf{G} \rangle \equiv (\gamma_F + \gamma_R)\,\gamma_P + \gamma_V \qquad [5.63]$$

It leads to:

$$(\mathbf{U}_m - {}^{\mathsf{t}}\langle \mathbf{G} \rangle)\,\mathbf{k}^* = \mathbf{c}_e \qquad [5.64]$$

where $\mathbf{c}_e = \boldsymbol{\nu}_0\,\hat{\mathbf{B}}^{-1}$, and:

$$\mathbf{k}^* = |\mathbf{B}\rangle\ \mathbf{c}_e \qquad \text{where} \quad |\mathbf{B}\rangle \equiv (\mathbf{U}_m - [\mathbf{G}])^{-1} \qquad [5.65]$$

Note that the characteristic equations for waste flows are given by:

$$\boldsymbol{\omega}_r = \gamma_R\,\gamma_P\,\mathbf{B} \qquad [5.66]$$

It can be proven that $\langle \mathbf{G} \rangle$ is an M-matrix (Berman and Plemmons 1979); therefore, equation [5.64] has a unique positive solution for the unit cost of the flows of the system. These results prove that the ECT is a particular approach of the structural theory of thermoeconomics, and the rules that assessed the exergy costs are rational rules derived from the definition of the exergy efficiency of the system processes. It also provides an alternative algorithm to compute the exergy cost of flows.

The coefficients k_{ij} of the characteristic matrix $\langle \mathbf{G} \rangle$ can be computed as:

$$k_{ij} = \begin{cases} x_i & i \in \mathcal{E}_l,\ j \in \mathcal{S}_l,\ l \in \mathcal{L}_u \\ x_i\,k_{u,l}\left(1 - \sum\limits_{e \in \mathcal{E}_h} x_e\right) & i \in \mathcal{E}_l,\ j \in \mathcal{S}_h,\ l \in \mathcal{F}_u,\ h \in \mathcal{P}_u \end{cases} \qquad [5.67]$$

where $k_{u,l} = F_{u,l}/P_u$, are the local unit consumptions, and x_i are the input ratios. It provides a general algorithm to build the matrix $\langle \mathbf{G} \rangle$, and then to compute the exergy cost of flows.

EXAMPLE.– The $\langle \mathbf{G} \rangle$ matrix of the TGAS plant, according to equation [5.67], is:

$$\langle \mathbf{G} \rangle = \begin{bmatrix} 1 & 2 & 3 & 3 & 3 & 4 & 4 & 5 & 6 & 9 \\ 3 & 3 & 4 & 6 & 7 & 5 & 8 & 9 & 2 & 1 \\ (1-x_2)\,k_1 & x_2 & 1 & k_3 & k_3 & 1 & k_4 & k_5 & k_2 & (1-x_2)\rho_5 \end{bmatrix}$$

where the value $x_2 \equiv B_2/B_3$ represents the exergy ratio between the air entering and the gases exiting the combustor, and $\rho_5 \equiv B_9/P_1$ is the ratio of waste gases per unit of product in the combustion chamber. The characteristic graph, including waste allocation equations (marked in red), are shown in Figure 5.12.

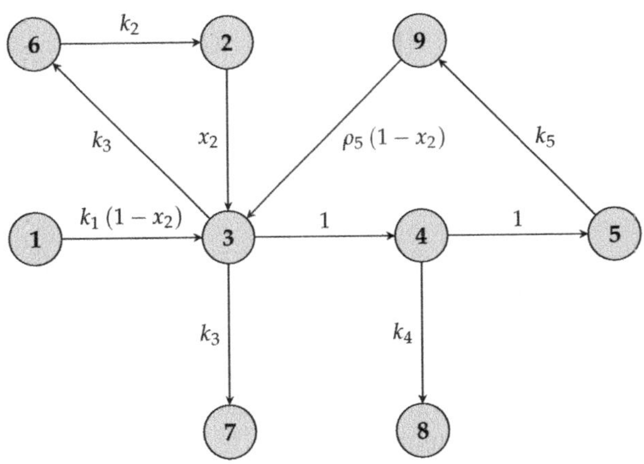

Figure 5.12. *Graph representation of TGAS characteristic equations from ECT, including waste allocation. For a color version of this figure, see www.iste.co.uk/feidt/advances.zip*

Therefore, the primal characteristic and the dual unit cost equations are as follows:

$$B_1 = (1-x_2)\,k_1 B_3 \qquad k_1^* = 1$$

$$B_2 = x_2\, B_3 \qquad k_2^* = k_2\, k_6^*$$

$$B_3 = B_4 + k_3(B_6 + B_7) \qquad k_3^* = x_2\, k_2^* + (1-x_2)k_1\, k_1^* + (1-x_2)\rho_5\, k_9^*$$

$$B_4 = B_5 + k_4\, B_8 \qquad k_4^* = k_3^*$$

$$B_5 = k_5\, B_9 \qquad k_5^* = k_4^*$$

$$B_6 = k_2\, B_2 \qquad k_6^* = k_3\, k_3^*$$
$$B_7 = \omega_7 \qquad k_7^* = k_3\, k_3^*$$
$$B_8 = \omega_8 \qquad k_8^* = k_4\, k_4^*$$
$$B_9 = (1 - x_2)\rho_5\, B_3 \qquad k_9^* = k_5\, k_5^*$$

These systems of equations can also be solved analytically or symbolically, following the scheme described in Figure 5.12. For example, the total fuel of the system can be expressed as a function of the local unit consumption of the processes, the input ratios and the exergy of the final products:

$$F_T = \frac{k_1\,(1 - x_2)\big[k_3\, B_7 + k_4\, B_8\big]}{1 - x_2\, k_2\, k_3 - (1 - x_2)\rho_5\, k_5}$$

5.4.3. *The fuel impact formula*

The fuel impact formula, due to the variation of the efficiency (or unit consumption of a process, can be obtained for equation [5.25]. In this case:

$$\frac{\partial G_0}{\partial k_u} = {}^{\mathrm{t}}\mathbf{k}^*\, \frac{\partial \langle \mathbf{G} \rangle}{\partial k_u} \mathbf{B}$$

Let introduce $\gamma_F^0 \equiv [\mathbf{F}]\,\hat{\mathbf{F}}^{-1}$, such that $\gamma_F = \gamma_F^0\,\hat{\mathbf{k}}$ and, $k_F^* = k^*\,\gamma_F^0$, therefore:

$$\frac{\partial G_0}{\partial k_u} = {}^{\mathrm{t}}\mathbf{k}^* \gamma_F^0\, \gamma_P \mathbf{B} = {}^{\mathrm{t}}\mathbf{k}_F^*\, \mathbf{P},$$

which coincides with the fuel impact formula [5.8] proposed by the ECT:

$$\mathcal{A}_{T,u} = \frac{\partial G_0}{\partial k_u} \Delta k_u = k_{F,u}^*\, P_u\, \Delta k_u \qquad [5.68]$$

It means that the additional resources consumption due to a relative variation of the efficiency of a local process Δk_u is proportional to the cost of its resources. For this reason, the calculation of the internal flows of a plant is crucial to understanding its behavior.

5.4.4. *Cost decomposition*

The exergy cost of internal flows calculated by equation [5.64] contains information both from the productive and dissipative structure of the system. To get a better understanding of the cost formation process, it is necessary to break down the cost into the contribution of the internal irreversibilities k^*_{IN} and waste k^*_{EX} or external irreversibilities.

The characteristic matrix could be decomposed as $\langle\mathbf{G}\rangle = \langle\mathbf{G}\rangle_{\mathrm{IN}} + \langle\mathbf{G}\rangle_{\mathrm{EX}}$, where:

$$\langle\mathbf{G}\rangle_{\mathrm{IN}} = \gamma_V + \gamma_F\gamma_P$$
$$\langle\mathbf{G}\rangle_{\mathrm{EX}} = \gamma_R\gamma_P$$

Considering the base cost operator, without considering waste allocation:

$$|\mathbf{B}\rangle_{\mathrm{IN}} \equiv \left(\mathbf{U}_m - \langle\mathbf{G}\rangle_{\mathrm{IN}}\right)^{-1}$$

therefore, the cost due only to internal irreversibilities is given by:

$$^t\mathbf{k}^*_{in} = {}^t\mathbf{c}_e\,|\mathbf{B}\rangle_{\mathrm{IN}} \qquad [5.69]$$

The characteristic matrix of waste could be written as $\langle\mathbf{G}\rangle_{\mathrm{EX}} = \boldsymbol{\delta}_R\,\boldsymbol{\rho}$, where $\boldsymbol{\delta}_R$ is a $(n \times r)$ matrix, whose columns are the unit vector $u_i = (0,\ldots,1,\ldots,0)$ corresponding to the dissipative component index i, and $\boldsymbol{\rho}$ are the active rows of the waste–product matrix $\gamma_R\gamma_P$.

Therefore, by applying the Woodbury formula (Hagger 1989), we can update the operator $|\mathbf{B}\rangle$, from the base production cost operator $|\mathbf{B}\rangle_{\mathrm{IN}}$, as follows:

$$|\mathbf{B}\rangle = |\mathbf{B}\rangle_{\mathrm{IN}}\left(\mathbf{U}_n + \boldsymbol{\delta}_R\mathbf{S}\boldsymbol{\rho}\,|\mathbf{B}\rangle_{\mathrm{IN}}\right)$$

where $\mathbf{S}$ is a $r(r \times r)$ matrix given by:

$$\mathbf{S} = \left(\mathbf{U}_r - \boldsymbol{\rho}\,|\mathbf{B}\rangle_{\mathrm{IN}}\,\boldsymbol{\delta}_R\right)^{-1}$$

Note that $r \ll n$, therefore the computation of matrix $\mathbf{S}$ is much faster than that of $|\mathbf{B}\rangle$.

Therefore, the unit cost of the flows due to waste is:

$$^t\mathbf{k}^*_{ex} = {}^t\mathbf{c}_r\,|\mathbf{B}\rangle_{\mathrm{IN}} \qquad [5.70]$$

where $\mathbf{c}_r$ is a $(m \times 1)$ vector which represents the unit cost of waste allocated to each flow, and is calculated as follows:

$$^t\mathbf{c}_r \equiv {}^t\mathbf{k}^* \, \gamma_R \, \gamma_P = {}^t\mathbf{k}^*_{in} \boldsymbol{\delta}_R \mathbf{S} \, \boldsymbol{\rho}$$

The unit cost of flows can be decomposed into the form:

$$\mathbf{k}^* = \mathbf{k}^*_{in} + \mathbf{k}^*_{ex} = {}^t \, |\mathbf{B}\rangle_{\text{IN}} \left(\mathbf{c}_e + \mathbf{c}_r\right) \quad [5.71]$$

EXAMPLE.– The exergy cost values of the flows of the TGAS plant are shown in Table 5.4. They are calculated from the exergy values of Table 5.1. To produce 1 kWh of combustion gases, we must spend 1.7116 kWh of natural gas. The combustion gases generate mechanical work in the turbine and steam in the HRSG. However, near 1 MW of natural gas is lost in the combustion gases exiting the plant. The cost of mechanical work is 1.8012 kW per kW of natural gas, and the cost of steam is 2.596 kW per kW of natural gas. It is due to the efficiency of HRSG being much lower than the efficiency of GTRB. The additional cost of waste flows per exergy unit (kWh) of gases produced in the combustor is less than 4% of the mechanical work.

Key	B(kW)	B^*(kW)	B^*_{in}(kW)	B^*_{ex}(kW)	k^*(J/J)	k^*_{in} (J/J)	k^*_{ex} (J/J)
NG	36620.0	36620.0	36620.0	0.0	1.0000	1.0000	0.0000
B2	9512.0	18416.9	17724.9	692.0	1.9362	1.8634	0.0728
B3	32990.0	56466.7	54344.9	2121.7	1.7116	1.6473	0.0643
B4	11707.0	20038.0	19285.1	752.9	1.7116	1.6473	0.0643
B5	835.3	1429.7	1376.0	53.7	1.7116	1.6473	0.0643
WC	10225.0	18416.9	17724.9	692.0	1.8012	1.7335	0.0677
WN	10000.0	18011.7	17334.9	676.8	1.8012	1.7335	0.0677
QV	7168.0	18608.3	17909.1	699.2	2.5960	2.4985	0.0975
QG	835.3	0.0	1376.0	53.7	1.7116	1.6473	0.0643

Table 5.4. *Exergy cost of flows for TGAS plant*

As shown in the introduction of the ECT, the cost of the flows is formed by the accumulation of the irreversibilities generated in producing them. Therefore, as we can determine the cost due to waste, we could also determine the cost due to internal irreversibilities and have a complete picture of the cost formation process.

From the construction of the characteristic matrix, it follows that:

$$ {}^t\mathbf{u}_m = {}^t\mathbf{c}_e + {}^t\mathbf{u}_n\,\gamma_P + {}^t\mathbf{u}_m\,\gamma_V $$
$$ {}^t\mathbf{u}_n = {}^t\mathbf{u}_m\,\gamma_F^0, $$

therefore, the cost of external resources $\mathbf{c}_e$ is given by:

$$ {}^t\mathbf{c}_e = {}^t\mathbf{u}_m\left(\mathbf{U}_m - \gamma_F^0\,\gamma_P - \gamma_V\right) $$

and substituting into [5.69], it follows:

$$ {}^t\mathbf{k}_{in}^* = {}^t\mathbf{u}_m + {}^t\left(\mathbf{k} - \mathbf{u}_n\right)\gamma_P\;|\mathbf{B}\rangle_{\mathrm{IN}} \qquad [5.72] $$

Let us define the *irreversibility matrix operator* $|\mathbf{I}\rangle$ $(n \times m)$ as:

$$ |\mathbf{I}\rangle \equiv \left(\hat{\mathbf{k}} - \hat{\mathbf{u}}_n\right)\gamma_P\;|\mathbf{B}\rangle_{\mathrm{IN}}, $$

which relates the system output with the local irreversibilities of the system processes, $\mathbf{I} = |\mathbf{I}\rangle\,\omega_0$, therefore equation [5.72] can be written as:

$$ {}^t\mathbf{k}_{in}^* = {}^t\mathbf{u}_m + {}^t\mathbf{u}_n\;|\mathbf{I}\rangle \qquad [5.73] $$

Equation [5.73], called the *irreversibility cost formula*, is essential to understand the process of cost formation. It shows that the exergy cost of a flow equals its exergy plus the sum of all exergy destroyed (irreversibility, losses and wastes) during the production process of that flow. In the above equation, ${}^t\left(\mathbf{k} - \mathbf{u}_n\right)\gamma_P$ represents the local exergy losses assessed to each productive group. Then, the cost operator $|\mathbf{B}\rangle_{\mathrm{IN}}$ distributes these values to the remaining processes of the system as additional resource consumption. The internal irreversibility reduction implies a decrease in production cost and savings in resources. Because the last term in [5.73] is positive, the exergy cost of every flow is always the greatest or equal to its exergy: $B_i^* \geq B_i$.

EXAMPLE.– Table 5.4.4 and Figure 5.13 show the cost formation process of flows for the TGAS plant. The graph shows that the most important irreversibility is generated in the combustor, affecting the cost of all flows. Meanwhile, the next highest irreversibility generated in the HRSG only affects the cost of its product flow QV. The cost of combustion gas flows B3, B4, B5 and QG is 1.7116 kW per kW of natural gas (NG). The values of the irreversibility matrix are shown in the rows of Table 5.4.4 corresponding to the production processes. The waste cost corresponds to the gray block in the graph, or the last row values (STACK) in the table.

Key	NG	B2	B3	B4	B5	WC	WN	QV	QG
COMB	0.0000	0.6687	0.5912	0.5912	0.5912	0.6221	0.6221	0.8966	0.5912
COMP	0.0000	0.1112	0.0321	0.0321	0.0321	0.0338	0.0338	0.0486	0.0321
GTRB	0.0000	0.0835	0.0241	0.0241	0.0241	0.0776	0.0776	0.0365	0.0241
HRSG	0.0000	0.0000	0.0000	0.0000	0.0000	0.0000	0.0000	0.5167	0.0000
STCK	0.0000	0.0728	0.0643	0.0643	0.0643	0.0677	0.0677	0.0975	0.0643
ENV	1.0000	1.0000	1.0000	1.0000	1.0000	1.0000	1.0000	1.0000	1.0000
Total	1.0000	1.9362	1.7116	1.7116	1.7116	1.8012	1.8012	2.5960	1.7116

Table 5.5. *Irreversibility Cost table for TGAS plant*

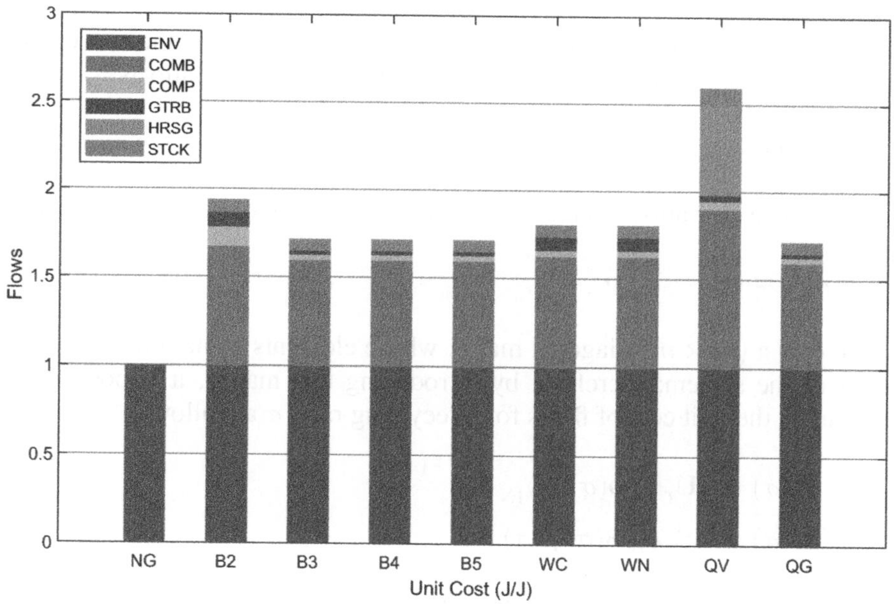

Figure 5.13. *Irreversibility cost graph for TGAS plant. For a color version of this figure, see www.iste.co.uk/feidt/advances.zip*

5.4.5. *Efficiency and recycling*

Whenever these waste streams have exergy, we can retrieve them in recovery units; this is the case for the heat recovery steam generator (HRSG) in the TGAS example. These recovery units are integrated into the plant and constitute productive units. Instead of expending exergy to eliminate the waste, their exergy is recovered to produce another by-product. In the case of the TGAS plant, the recovery unit produces saturated steam.

If we know the production costs due to external irreversibilities, we can estimate the potential savings from waste recycling. The production cost due to waste C_P^r indicates the maximum amount of resources saved if waste is recycled.

It is well known that not all waste can be saved for technical and economic reasons. Through the process integration and industrial symbiosis (Valero et al. 2013), other systems may convert this waste into resources. The ECT allows the cost of waste to be evaluated, accounting for resources savings and providing fair prices to waste becoming by-products.

We can now define the *recycling ratio* σ_i, as the percentage of waste i that can be reused or, in these conditions $\omega_{S,i} = \sigma_i \, \omega_{R,i}$. Therefore, the cost of fuel due to waste can be computed as a function of the recycling ratio as:

$$k_{ex}^*(\sigma) = {}^t\mathbf{c}_r(\sigma)\,|\mathbf{B}\rangle_{\text{IN}} \qquad [5.74]$$

The waste cost allocation matrix is a linear function of the recycling ratio:

$$\boldsymbol{\rho}(\sigma) = (\mathbf{U}_r - \hat{\boldsymbol{\sigma}})\,\boldsymbol{\rho}$$

where $\hat{\boldsymbol{\sigma}}$ is a $(m \times m)$ diagonal matrix whose elements contain the *recycling ratios* of the system; therefore, by introducing this matrix, it is possible to recalculate the unit cost of flows for a recycling ratio σ as follows:

$$\mathbf{S}(\sigma) = \left(\mathbf{U}_r - \boldsymbol{\rho}(\sigma)\,|\mathbf{B}\rangle_{\text{IN}}\,\boldsymbol{\delta}_R\right)^{-1}$$
$${}^t\mathbf{c}_r(\sigma) = {}^t\mathbf{k}_{in}^*\boldsymbol{\delta}_R\,\mathbf{S}(\sigma)\,\boldsymbol{\rho}(\sigma)$$

The *cost conservation* rule applies as the costing rule in a recovery unit. The exergy cost of final waste plus the obtained by-product is the cost of entering waste plus the fuel spent to diminish them. It reduces the cost of waste. Then, at the end of the analysis, only these waste costs add to the final cost of the plant product(s). Recovering the exergy of waste saves exergy and reduces the cost of products. The use of *recovery units* or *processes integration* to recover part of the exergy contained in waste streams is a fundamental practice for saving energy.

EXAMPLE.– Consider the TGAS plant again, and suppose there is not an HRSG to recover part of the escape gases of the turbine. In this situation, the cost of electricity will be 3.662 kW of natural gas for kW of electricity produced, and moreover, we have lost 11,707 kW in hot gases leaving the gas turbine.

Using the formulation introduced in this section, it is possible to compute the production cost of electricity and combustion gases depending on the recycling ratios, as it is shown in Figure 5.14. If the exhaust gases from the turbine were not recovered, the unit cost of electricity would be 3.662, almost twice as much as if an HRSG is added to the turbine, and 92% of the exhaust gases are recovered.

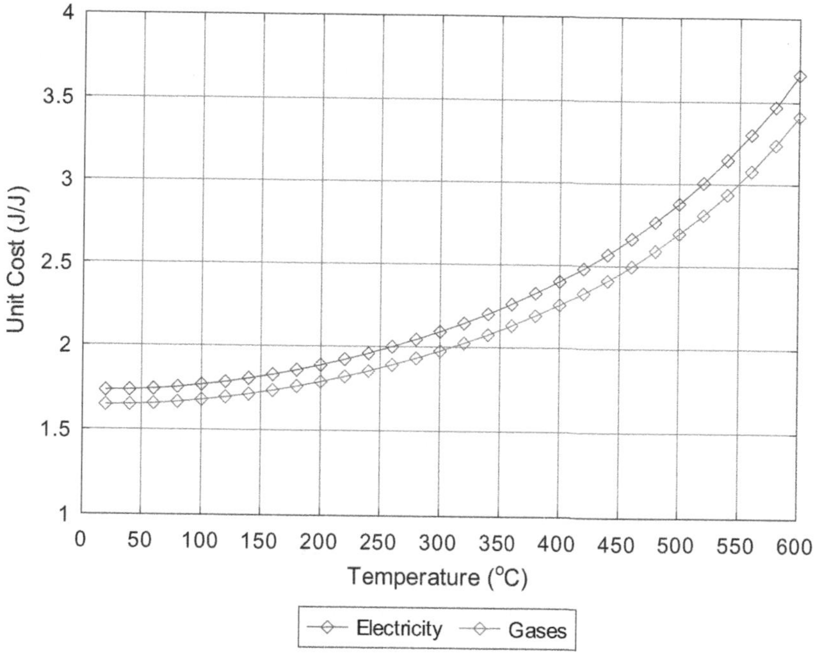

Figure 5.14. *Recycling saving graph for TGAS plant. For a color version of this figure, see www.iste.co.uk/feidt/advances.zip*

The unit cost of electricity can be expressed as a function of the recycling ratio:

$$k_7^*(\sigma) = \frac{(1 - x_2)\, k_1\, k_3}{1 - k_2\, k_3\, x_2 - \rho_5\, (1 - x_2)(1 - \sigma)}$$

In general, the unit cost as a function of the recycling waste ratios has the form:

$$k^*(\sigma) = \frac{a}{b + c\sigma}, \quad \sigma \in [0, 1], \quad a, b, c > 0 \qquad [5.75]$$

where a, b, c are coefficients that depend on the local unit consumption of the system components.

This function is nonlinear, decreasing and convex (see Figure 5.14). It means that unit consumption decreases $a/b > a/(b + c)$ and exergy savings increase when the proportion of waste recycling increases, but not in a linear way. Because the function is convex, the exergy saving at the beginning of the recycling process is very significant, but as we approach its theoretical limit, that is, its reference environment, the energy savings are smaller and smaller, until a practical minimum is reached that could be called the *pinch point of recoverability*. For example, by adding other recovery processes, such as absorption units, or improving the HRSG, we could recycle up to 98%, but the total fuel consumption only decreases by 2% of natural gas consumption.

5.5. Conclusion

The aim of this chapter has been twofold. On the one hand, it describes the structural theory as the conceptual basis of *circular* thermoeconomics by delving into the physical behavior of productive systems in which energy plays a relevant role. On the other hand, it analyzes the corollaries of this theory in those social and industrial initiatives, which treat waste as a complementary resource solution to the decarbonization of society. Consequently, we divide the conclusions into two parts: the first, those related to structural theory and the second, those applicable to the circular economy and industrial symbiosis.

5.5.1. *The structural theory*

The purpose of thermoeconomics is to obtain a coherent and significant set of costs in a given energy structure. That is because its applications are directed toward the optimization and diagnosis of productive structures. Therefore, the costs of internal streams are quite important to analyze the performance of the system. Consequently, internal costs ought to be as sensitive as possible to system degradation. A theory that would provide exact costs would be the objective of thermoeconomic diagnosis, that is, how to assess the impact on additional resources due to small variations in the behavior of the plant.

The structural theory solves and explains several important questions of the conventional exergy costing theories. First, its results do not depend on a disaggregation scheme. For example, if the TGAS system is aggregated into a single component, the physical behavior of the system remains (if linearized) the same under equations [5.16] and [5.26]. However, neither the rules F or P can be applied between streams #7 or #8. In other words, k_7 is not equal to k_8, nor to k_1.

Therefore, the theory provides many hints on exploring the cost relationships and suggests that greater disaggregation schemes are needed to reconcile genuine product and waste cost formation with its internal economy. The key idea originates from the difference between the characteristic equation and the efficiency definition. The former reflects the physical behavior of the component, while the second is an identity. The similarity is evident; however, the more the analyst aggregates, the more the physical behavior of the system deviates from the aggregated efficiency definition. In fact, the structural theory explains why some definitions of efficiency reflect the physical behavior of a piece of equipment better than others. The physical cost, stripped of economic arguments, must sensitively reflect the physical performance of the flows that interact in a system.

These ideas are key to understanding the connection between thermodynamics and economics. In the same way, they serve to find out which is the optimal disaggregation of a system to perform a meaningful thermoeconomic analysis. It must be taken into account that, at least in theory, it would be possible to evaluate, empirically or with a simulator, the variation of the total resources of a plant when equipment degrades under some predetermined conditions. Both the local entropy increase (or exergy decrease) and the overall additional resources are measurable and free from analyst selections. The difference is clear, performing an experiment in a real plant under prefixed conditions is almost unattainable. Meanwhile, a simulator may provide hardly interpretable results.

By contrast, thermoeconomic analysis allows for understanding and quantifying induced malfunctions and side effects of the relationships between local and overall effects. A simulator describes the behavior of a system using physical magnitudes, such as masses, pressures, temperatures, efficiencies, ratios, etc., as independent variables. At the same time, a thermo-economic analysis focuses on products and irreversibilities both internal and external, measured in the same units as the resources entering the plant, i.e. in kWh.

The structural theory, first presented in Valero et al. (1992a, 1993), formally relates to the input-output theory (Leontief 1970); however, the characteristic equation makes the difference because it deeply connects the thermodynamic behavior of an energy system to costs such as in a transparent box. This theory could be considered a bridge between classical economics and physics. Meanwhile, the Exergy Cost Theory (Lozano and Valero 1993) provides reasonable costs and can be used instead, provided that the caveats indicated in this article are taken into account.

5.5.2. *Circular economy and industrial symbiosis*

Thermoeconomics is a science derived from thermodynamics, which applies to all known physical systems. Therefore, what is valid for thermodynamics is also true for thermoeconomics. Consequently, thermoeconomics allows us to generalize the idea of waste as any useless mass and/or energy flow. In fact, waste is an exergy flow that will dissipate sooner or later until it reaches equilibrium with the environment. It is simply external irreversibility that will generate entropy outside the system under analysis. Hence, the cost of waste is formed by the exergy of the resources used to produce, transfer and dispose of it within the system. It can be measured in exergy units (kWh), but it should not be forgotten that the recovery of waste always involves the production of new waste.

According to this definition of waste, the circular economy ignores the fact that cycles can never be closed because some of the waste can never be reused. In reality, technology focuses on the recovery of the most valuable component, leaving behind a waste that is less and less recyclable. Furthermore, it mainly deals with solid waste and forgets to take into account gas emissions, reagents or other indirect costs associated with reprocessing. This is an important idea, just as every product has an impact on greenhouse gases, as does every reprocessing of a waste product. The theory developed here allows us to obtain this impact (Agudelo et al. 2013).

As a result, there can be no “Zero Waste” solution. What we could achieve at most, as a thermodynamic limit, is a zero-waste exergy system. Although physics does not prevent it, this would be an extreme case. This process would consist of extracting all the exergy from the waste. It is a series of reversible processes that would return each chemical element to nature in composition and concentration equal to that of the reference state to calculate exergy, i.e. Thanatia (Valero and Valero-Delgado 2014). The limit imposed by the prevailing technology is far removed from the thermodynamic one and even further from the economic one, or even more from the legal and administrative reality.

Besides, there is a point where total recycling is unattainable. As the intensive properties of waste, such as its composition, concentration, pressure or temperature approach the corresponding properties of the reference environment, the savings obtained are smaller and smaller, until a practical minimum is reached that could be called the *pinch point of recoverability*. Moreover, as we approach these technical limits, the waste's usefulness decreases compared to what nature provides for free. Considering the entropic behavior of the separation processes, we can say that the inability of our technology increases exponentially as we reach the pinch point of each intensive property of the waste.

Let us take a look at the combustion gases of a process. If the gases have a high temperature (high specific exergy), the capacity to extract useful work is high. It is the case of a heat recovery boiler or an economizer. As the gases are approaching the environment temperature, it is more difficult to obtain useful work from them; sometimes depending on the chemical composition of these gases, it is technically impossible due, for example, to corrosion problems. The case of a Rankine cycle condenser is similar; as the condensation temperature approaches the ambient temperature, it is more challenging to extract useful work from it.

Despite this, nature is incredibly more efficient than our technology. All living beings produce waste while they are alive, and they are waste once they are dead, they are recycled to an overwhelming extent driven by solar energy. We can, therefore, only humbly learn from nature (Benyus 1997).

The proposed methodology, circular thermoeconomics, based on the exergy cost theory and its generalization, the structural theory, allows the accounting of physical or thermodynamic costs, both of functional products and waste produced in parallel, and also allows the determination of which part of the production costs is due to internal irreversibilities of the plant, and which part is due to the waste produced or external irreversibilities. Each of these internal and external irreversibilities can be reduced to certain limits. Internal irreversibilities, by improving process efficiency or changing technology, are always under certain technical limits. Waste can be reduced by recycling, but internal irreversibilities are also the cause of waste, inducing new costs due to waste relocation and reduction. Therefore, the reduction of irreversibilities, both internal and external, are closely related.

The word “recycling” is used interchangeably by thermoeconomics, circular economy and industrial symbiosis. We may wonder if it means the same thing in all three fields of activity. Recycling means efficiency in the use of resources. Therefore, what changes is the type of system being studied. In general, the external recovery of any solid waste is the subject of the circular economy, while its system to be analyzed is the recovery plant. Industrial symbiosis refers to the process in which a waste or by-product from one plant is reused in other plants. In this case, the system under study is the set of interacting plants. Thermoeconomics covers the general case, no matter how complex the subsystems and flow interconnections are.

If we look at equation [5.75], every time a σ percentage of waste is recycled, the efficiency of the system as a whole improves all the more as σ increases. However, when the proportion of waste recycling increases, efficiency does not increase linearly. This is because the function is convex, that is, the exergy savings at the beginning of the recycling process are quite remarkable, however, as we approach its theoretical limit – the reference environment – the exergy saving nullifies. It is another mathematical demonstration of the uselessness of enforcing the recycling rate to the limit and a plea for the spiral economy. An additional increase in recycling does not correspond to the same increase in efficiency gain. As a matter of fact, this is the law of diminishing returns.

If waste leaves the boundaries of a plant, two things can happen:

– It can dissipate into the environment, but its formation costs must be accounted for, identifying its origin and internalizing those costs to the processes that have produced it. In other words, for a fully responsible industry, the company will bear the cost of its waste and endorse this cost on the final product(s).

– Part of this waste can be reused in other processes or plants, in which case the theory allows for the calculation of its formation costs, and will provide an objective cost to the company as a basis for discussion with other plants interested in using it.

The so-called circular economy cannot circumvent the second law of thermodynamics, nor can it proclaim that everything can be recycled indefinitely. It takes into account neither the time that nature has at its disposal nor the production of waste that recycling entails. The idea of circular economy is a metaphor used to relieve consumers of responsibility. They are led to believe that, by separating their waste into its corresponding containers, they have already recycled. It should, in consideration of the law of entropy, be renamed spiral economy. All the more so as in most cases today, only a first cycle can be achieved by producing down-cycled products.

A spiral economy accepts the impossibility of completely closing the cycles, but instead allows us to analyze how far we are from achieving this. This makes it possible to define the efficiency of circularity as the exergy gained divided by the exergy expended in reprocessing. Moreover, it allows us to have a comprehensive view of the degradation process of the recycled material, as well as to identify a radius (always decreasing) of the spiral related to the valuable residence time of the material in society.

It is often repeated that in every threat, there are opportunities. However, there are many materials that cannot be recycled, or if they are, they degrade and cannot be used indefinitely. For example, the thousands of chemicals produced each year, plastics, paper, wood, tyres, special alloys, etc. These opportunities are not obvious, because we are faced with the grim and most limiting physical law in the material world: the second law of thermodynamics. In fact, to recover highly non-recyclable products, we would have to resort to processes that are the opposite of those used in their synthesis, in which the energy cost of separation is too high, and which in turn will produce new solid, liquid and gaseous waste.

In this endeavor, citizen awareness and laws must change, and moderation in consumption must be imposed that accepts the biophysical limits of the planet. However, it is engineering that must play a central role. New processes will be created to wisely repair what is broken (kintsugi), or recover and convert waste into resources. Industrial symbiosis will be generalized so that waste from one plant becomes a resource for another (cascading waste into feedstock). Also, separation technologies will be developed to minimize its own product-to-be-waste (eco-efficiency). This will be achieved by exploiting their energy and chemical potential, improving the efficiency of processes, recycling and converting them into new raw materials, as nature does (biomimicry and renewable energies). This is a field of research that requires much more new knowledge. It is even discouraging to see how current trends in science celebrate the achievement of new properties of increasingly entropic materials, that is, high entropy alloys (Lim 2016), and yet there is a huge gap in research into the separation of their components. Separation technologies are considered a minor activity within the most distinguished research. It is not easy to find advances in high impact journals. Even though the name circular economy is an oxymoron that strictly violates the second law of thermodynamics, it is a social need that must be developed urgently because it helps to save natural resources.

In summary, circular thermoeconomics, presented here, is a general procedure based on the structural theory of thermoeconomics, which can be used for waste cost accounting both in the circular economy and industrial symbiosis.

5.6. References

Agudelo A., Valero, A., Usón S. (2013). The fossil trace of CO_2 emissions in multi-fuel energy systems. *Energy*, 58, 236–246.

Ashby, A., Callegaro, A.M., Adeyeye, K., Granados, M. (2019). The spiral economy: A socially progressive circular economy model? In *Sustainable Development Goals and Sustainable Supply Chains in the Post-global Economy*, Yakovleva, N., Frei, R., Rama Murthy, S. (eds). Springer, Cham, Switzerland.

Bejan, A., Tsatsaronis, G., Moran, M.J. (1996). *Thermal Design and Optimization*. Wiley, New York, USA.

Benyus, J.M. (1997). *Biomimicry: Innovation Inspired by Nature*. William Morrow and Company, New York, USA.

Berman, A. and Plemmons, R.J. (1994). *Nonnegative Matrices in the Mathematical Sciences*. Society for Industrial and Applied Mathematics, Philadelphia, PA, USA.

Bertalanffy, L.V. (1968). *General System Theory*. George Braziller Inc., New York, USA.

Cecchin, A., Salomone, R., Deutz, P., Raggi, A., Cutaia, L. (2020). Relating industrial symbiosis and circular economy to the sustainable development debate. In *Industrial Symbiosis for the Circular Economy: Operational Experiences, Best Practices and Obstacles to a Collaborative Business Approach*, Salomone, R., Cecchin, A., Deutz, P., Raggi, A., Cutaia, L. (eds). Springer International Publishing, Cham, Switzerland.

Da Silva, J.A.M., Santos, J.J., Carvalho, M., De Oliveira, S. (2017). On the thermoeconomic and LCA methods for waste and fuel allocation in multiproduct systems. *Energy*, 127, 775–785.

El-Haggar, S.M. (2007). Cleaner production. In *Sustainable Industrial Design and Waste Management*, El-Haggar, S.M. (ed.). Academic Press, Oxford, UK.

Ellen MacArthur Foundation (2013). Towards the circular economy. Vol. 2: Oportunities for the consumer goods sector. Technical report, Ellen MacArthur Foundation, Cowes, UK.

Frangopoulos, C. (1987). Thermo-economic functional analysis and optimization. *Energy*, 12(7), 563–571.

Gonzalez, A., Sala, J., Flores, I., Lopez, L. (2003). Application of thermoeconomics to the allocation of environmental loads in the life cycle assessment of cogeneration plants. *Energy*, 28(6), 557–574.

Graedel, T. and Allenby, B. (1995). *Industrial Ecology*. Prentice Hall, Upper Saddle River, NJ, USA.

Hager, W.W. (1989). Updating the inverse of a matrix. *SIAM Review* 31, 221–239.

Keshavarzian, S., Rocco, M.V., Gardumi, F., Colombo, E. (2017). Practical approaches for applying thermoeconomic analysis to energy conversion systems: Benchmarking and comparative application. *Energy Conversion and Management*, 150, 532–544.

Kotas, T. (1985). *The Exergy Method of Thermal Plant Analysis*. Butterworths, London, UK.

Kumar, R. (2017). A critical review on energy, exergy, exergoeconomic and economic (4-E) analysis of thermal power plants. *Engineering Science and Technology, an International Journal*, 20(1), 283–292.

Lazzaretto, A. and Tsatsaronis, G. (2006). SPECO: A systematic and general methodology for calculating efficiencies and costs in thermal systems. *Energy*, 31(8), 1257–1289.

Leontief, W. (1970). Environmental repercussions and the economicstructure: An input–output approach. *Review of Economics and Statistics*, 52(3), 262–271.

Lim, X. (2016). Metal mixology. Stronger, tougher, stretchier: With a simple new recipe, metallurgists are creating a generation of alloys with remarkable properties. *Nature*, 553, 307–308.

Lozano, M. and Valero, A. (1993). Theory of the exergetic cost. *Energy*, 18(9), 939–960.

McDonough, W. and Braungart, M. (2003). *Cradle to Cradle: Remaking the Way We Make Things*. North Point Press, New York, USA.

Morris, D.R. and Szargut, J. (1986). Standard chemical exergy of some elements and compounds on the earth planet. *Energy*, 11, 733–755.

Querol, E., Gonzalez-Regueral, B., Perez-Benedito, J.L. (2013). *Practical Approach to Exergy and Thermoeconomic Analyses of Industrial Processes*. Springer, London, UK.

Sciubba, E. (2003). Extended exergy accounting applied to energy recovery from waste: The concept of total recycling. *Energy*, 28(13), 1315–1334.

Seyyedi, S.M., Ajam, H., Farahat, S. (2010). A new criterion for the allocation of residues cost in exergoeconomic analysis of energy systems. *Energy*, 35(8), 3474–3482.

von Spakovsky, M. and Evans, R. (1988). Optimal design and performance of components in thermal systems. *American Society of Mechanical Engineers*, 6, 65–81.

Szargut, J., Morris, D.R., Steward, F. (1988). *Exergy Analysis of Thermal, Chemical, and Metallurgical Processes*. Hemisphere, New York, USA.

Torres, C. and Valero, A. (2012). The fuel impact formula revisited. *Proceedings of the 25th International Conference on Efficiency, Cost, Optimization and Simulation of Energy Conversion Systems and Processes, ECOS 2012*, Perugia, Italy.

Torres, C. and Valero, A. (2021). The exergy cost theory revisited. *Energies*, 14, 1594.

Torres, C., Valero, A., Perez, E. (2007). Guidelines to developing software for thermoeconomic analysis of energy systems. Part I: The thermoeconomic model. *Proceedings of the 20th International Conference on Efficiency, Cost, Optimization, Simulation and Environmental Impact of Energy Systems, ECOS 2007*, Padua, Italy.

Torres, C., Valero, A., Rangel, V., Zaleta, A. (2008). On the cost formation process of the residues. *Energy*, 33(2), 144–152.

Torres, C., Usón, S., Valero, A., Valero-Delgado, A. (2010). Thermoeconomics meets industrial ecology. *Proceedings of the 23rd International Conference on Efficiency, Cost, Optimization, Simulation, and Environmental Impact of Energy Systems, ECOS 2010*, Krakow, Poland.

Tsatsaronis, G. and Winhold, M. (1985). Exergoeconomic analysis and evaluation of energy-conversion plants I: A new general methodology. *Energy*, 10(1), 69–80.

Valero, A. and Torres, C. (2009). Thermoeconomic analysis. In *Exergy, Energy System Analysis and Optimization*, vol. 2, Frangopoulos, C. (ed.). Eolss Publishers, Oxford, UK.

Valero, A. and Torres, C. (2020). Relative free energy function and structural theory of thermoeconomics. *Energies*, 13(8), 2024.

Valero, A. and Valero-Delgado, A. (2014). *Thanatia: The Destiny of the Earth's Mineral Resources*. World Scientific, London, UK.

Valero, A., Lozano, M., Muñoz, M. (1986). A general theory of exergy saving. Part I: On the exergetic cost. *Computer-aided Engineering and Energy Systems: Second Law Analysis and Modelling*, 3, 1–8.

Valero, A., Guallar, J., Muñoz, M. (1987). The exergetic cost and related concepts: An application to a simple cogeneration plant. *Proceedings of Fourth International Symposium on Second Law Analysis of Thermal Systems*, ASME, I00236, 123–130.

Valero, A., Torres, C., Lozano, M. (1989). On the unification of thermoeconomic theories. *American Society of Mechanical Engineers*, 124, 63–74.

Valero, A., Torres, C., Serra, L. (1992). General theory of thermoeconomics. Part I: Structural analysis. *Proceedings of ECOS '92 International Symposium*, Zaragoza, Spain.

Valero, A., Serra, L., Lozano, M. (1993). Structural theory of thermoeconomics. *American Society of Mechanical Engineers*, 30, 189–198.

Valero, A., Lozano, M.A., Serra, L., Tsatsaronis, G., Pisa, J., Frangopoulos, C., von Spakovsky, M.R. (1994). CGAM problem: Definition and conventional solution. *Energy*, 19(3), 279–286.

Valero, A., Correas, L., Lazzaretto, A., Rangel, V., Reini, M., Taccani, R., Toffolo, A., Verda, V., Zaleta, A. (2004). Thermoeconomic philosophy applied to operating analysis and diagnosis of energy utilty systems. *Int. J. Thermodynamics*, 7(2), 33–39.

Valero, A., Uson, S., Torres, C., Valero, A., Agudelo, A., Costa, J. (2013). Thermoeconomic tools for the analysis of eco-industrial parks. *Energy*, 62, 62–72.

6

The Relative Free Energy Function: A New Approach to Thermoeconomic Diagnosis

Antonio VALERO-CAPILLA and César TORRES
CIRCE Institute, University of Zaragoza, Spain

This chapter explores the advantages of using relative free energy instead of exergy to build a mathematical theory of thermodynamic costs to diagnose malfunctions in thermal systems. This theory is based on the definition of a linearized characteristic equation that represents the physical behavior of each component. As we saw in the previous chapter when describing the structural theory, the primal is the physical structure of the system linked by energy interrelationships, and its derivatives are the unit costs and consumptions. In this chapter, we see that another important consequence of the theory is a new thermodynamic function called the *relative free energy*, and a new parameter called *deterioration temperature* due to a component's deterioration cause, characterized by a h-s thermodynamic trajectory related to the effects on the exiting stream. The relative free energy function allows for an exact relationship between the amount of used resources and the increase in entropy generation caused by the deterioration path of the component. This function allows the obtaining of an appropriate characteristic equation for a turbine and a new definition of efficiency that does not depend on the environment temperature but on its deterioration temperature. Also, costing with relative free energy instead of exergy may

Advances in Thermodynamics and Circular Thermoeconomics,
coordinated by Michael FEIDT and Antonio VALERO-CAPILLA. © ISTE Ltd 2023.

open a new path for more precise and straightforward assessments of component deteriorations.

6.1. Introduction

The aim of this chapter is challenging since it questions all the concepts that have been traditionally used in thermoeconomics, also called exergoeconomics by other authors (Tsatsaronis and Winhold 1985). There is a consensus among practitioners that exergy is the thermodynamic property that best distributes production costs among various streams, which then helps to establish the economic costs of the streams which interact in the energy system. However, the exergy calculation has the shortcoming of depending on the reference environment applied, which in turn is an arbitrary selection of the user and not a physical behavior of the system under analysis. If this is the case, the concept of exergy efficiency, although more precise than the first law efficiency, still depends on the reference environment. This obviously applies also to its inverse, the unit exergy consumption, which is the precursor of the theory of the exergy cost (Valero et al. 1986; Lozano and Valero 1993). Efficiency and costing are in effect, two sides of the same coin.[1]

The exergy cost theory (ECT) is simple to understand and easy to implement computationally, and as a result, its usage has taken off since 1986, when it was first published. A year earlier, Tsatsaronis proposed to identify all the currents of an energy system in fuels and products and thereby generate the exergoeconomic costs. The ECT facilitated the task by proposing general fuel-product cost allocation rules that made the costing of energy systems a quasi-routine matter. But the problem of selecting a reference environment remains and there might a consensus regarding which is the most suitable. However, convergent opinions are not mathematical evidence. Moreover, the word thermoeconomics adds an enormous responsibility to the researcher, as it implicitly makes a fundamental connection between thermodynamics and economics.

Historically, many economists, especially the so-called ecological economists, have approached thermodynamics. The work of Martinez-Alier (1987) extensively describes the early connections between economists and energeticists. Perhaps the most rigorous work was that from

1. This chapter is partially based on the paper *Relative Free Energy Function and Structural Theory of Thermoeconomics* (Valero and Torres 2020).

Georgescu-Roegen (1971). He already discarded energy as a numeraire, pointing to entropy instead. He even went beyond by formulating the fourth law of thermodynamics, as the impossibility of complete recycling of materials, drawing a parallel with the second law as the impossibility of complete recycling of thermal energy. From there on, ecological economists have tried to use entropy more as a metaphor than as a tool that quantifies decisions.

On the other hand, thermodynamicists, mainly engineers, developed the concept of exergy. It is not the aim of this chapter to describe the history (see Sciubba and Wall 2007), but to outline the conceptual roots of thermoeconomics. Exergy measures the theoretical ability to do work, therefore exergy efficiency will more precisely measure the performance of a machine. Yet the inverse of the efficiency describes the amount of resources consumed to obtain a product. And this is actually the definition of the physical cost, which in turn stands on the threshold of the economics. This is so because if we value resources in monetary units and add other costs such as equipment depreciation, maintenance and other expenses, we obtain the monetary costs of production not only of the final products but of the intermediate products (Tribus and Evans 1962; El-Sayed and Evans 1970; Tsatsaronis and Winhold 1985). This opened the door to compare the costs of irreversibilities against the costs of operation and maintenance, which in turn naturally lead to propose improvements in the design of industrial processes. This was a paramount technological breakthrough.

Arguably, practicality and not mathematical precision is what is sought in exergoeconomic cost analysis because the cost is measured in monetary units. That said, the fluctuation of money is alien to the process designer, since it depends on external conditions that the practitioner cannot control. Consequently, exergoeconomics is a useful tool in the design of industrial processes, but it does not strive for the same precision associated with physical sciences. Its results depend both on the variability of money and on the measurement of irreversibilities with respect to the chosen reference state.

The question is whether to go back to the concept of exergy cost, which only measures how the physical resources consumed by the production plant are distributed. In so doing, we at least avoid currency fluctuations and focus on the diagnosis of malfunctions that occur in all industrial processes (Valero et al. 2004a). This opened a new technological path because no energy system deteriorates for a single cause, but for several at the same time with different intensities and with chains of causation that induce new malfunctions and dysfunctions. Thermoeconomic diagnosis, using exergy costs, can be applied online and is very useful when analyzing very complex plants. The disaggregation of any industrial energy system into subsystems allows us to understand the process of the formation of physical costs, predict failures and it may keep the plant in the optimal operating conditions. Yet the cost of malfunctions is still dependent on the reference state of the plant, which can be taken as fixed or variable depending on the external environmental conditions.

In other words, exergy is useful as an industrial diagnosis tool, equipment malfunction is essentially due to deterioration of the equipment itself, not due to the definition of the reference environment. In short, exergetic diagnosis is not a precise and objective physical theory, but a very useful tool, agreed with by practitioners. The question now is: is there an exergy-like type function that could provide diagnoses which are independent from the reference state? This would mean that the environmental conditions would, for sure, influence the industrial system, not because of an arbitrary designer's selection, but through the environment's own natural interaction with the process.

If this were possible, we could give a rigorous answer not only to the thermodynamics of energy systems but to the economics itself, and all without straying from the second law. Linking the concept of cost with thermodynamics is to bridge the gap between physics and economics. It would be a crucial conceptual step forward sought by economists, something like the Holy Grail of economics. Indeed, while economists have come a long way in mathematical applications, they have never been able to link their laws to any physical law, except from metaphors and analogies. In this way, the conventional separation between hard and soft sciences would be closer to a scale of hardness than to the hard/soft dichotomy.

This chapter expands upon an early keynote lecture at Nancy University (Valero 2018), and its objective is to come closer to a solution to this problem and investigate the concept of relative free energy, already proposed by Valero and Lozano (1992), but with important precursors such as Baehr (2005) and Alefeld (1988). The ideas developed here are a consequence of the structural theory (Valero et al. 1992a) explained in the previous chapter. There, we analyzed whether or not the decisions made by the process analyst are or are not supported by mathematical logic. These decisions are mainly the cost allocation rules, the level of aggregation, the fuel/product/waste definitions and, of course, the exergy function and the reference state. This now allows us to discern the mathematical reasons for reasonable assumptions and to move toward a new thermodynamic function: *the relative free energy function.*

6.2. Drawbacks of exergy

One important result of the structural theory is that any energy function of the type $E = H - T_x S$ may be used to assess different sets of costs. Thus, why should exergy that uses a fixed T_0 instead of a variable T_x be considered? The most important objective of this study is to question the exergy directly to assess costs.

Mythologizing exergy is not convenient, it only measures the number of times that a product is potentially equivalent to another. Therefore, the kilograms of natural gas; mass of steam at a certain temperature and pressure; and kinetic, magnetic, or mechanical energies are possible units for measuring exergy. Certainly, it does not reflect the product value. A broken glass has practically the same exergy as a new one, and the exergy of gold at T_0 is zero as taken by reference. Moreover, everybody would prefer a 1 g diamond compared to a 1 g piece of pure graphite, despite them practically having the same exergy.

In addition, the manufacture of a continuous and broken thread costs the same exergy, although the latter would not be saleable. Exergy reduces a probably irreducible object to a single numerical value. The color, taste, form, texture, or artistic impression (i.e. properties that human beings value) have no distinguishable exergy. The exergy and exergy cost are only a measure of reality, which should not supplant other possible analyses.

Can the concept of value be assessed by thermodynamics? Does it make sense to go beyond the exergy cost when it comes to knowing the cost of the objects? It has been claimed many times that exergy is a measure of the quality of things, which is false; it is rather a measure of the energy imbalance that a system has with respect to a reference environment. It is a measure of the thermodynamic quality of energy and not of objects. Moreover, the quality specification is much more complex. A dose of carbon monoxide may have the same exergy as a certain amount of food; however, everybody knows which one to ingest or burn in a boiler. The same food can present a disgusting or an extraordinary aspect and contain the same amount of exergy. Based on the many possible examples, it seems strange to confuse exergy with quality.

Assessing thermodynamic quality requires many specifications such as pressure, temperature, composition, height and speed. Quality also depends on the context, whereas exergy does not. In general, thermodynamic equivalence is not a measure of the equivalence of values. Even circumstances and the moment make a thing seem more or less valuable. Researchers should distance themselves from neo-energeticism (or even better *exergeticism*), which regards exergy rather as a measure of the value of objects than as their monetary value. The exergy and exergy costs provide complementary information on the human footprint on Earth; ignoring them is as dangerous as extolling them as a unique instrument for managing the natural resources and environment.

Furthermore, the second law has not exhausted its message regarding exergy. The following equation:

$$F - P = I + R \qquad [6.1]$$

identifies the product, the waste and what is (or is not) a usable resource. Certainly, every real-world process experiences irreversibility. The Gouy–Stodola theorem relates the entropy generation to irreversibility:

$$I = T_0 S_g \qquad [6.2]$$

Regardless of the way the quality of every stream, product or service is defined (generalized exergy function), it is easier to identify the quantity. Thus, the magnitude of a property X is always a product of its quantity q_x multiplied by its specific quality x:

$$X = q_x \cdot x \qquad [6.3]$$

It would be interesting to determine the amount of resources required to compensate for a deterioration of any component in a constant production for a set of quality specifications for each flow interacting in a system.

6.3. The relative free energy function

Before generalizing the problem, a simple case should be analyzed: for example, the expanding flow of a turbine produces work on the shaft. In an (h, s) diagram (Figure 6.1), the actual process evolves from state 1 to 2.

The energy and entropy balances of the process (first and second laws) are as follows:

$$W = m\,(h_1 - h_2) \qquad [6.4]$$

$$S_g = m\,(s_2 - s_1) \qquad [6.5]$$

If the turbine degrades, at a rate equal to the specifications of the inflow #1, an increase in the entropy generation evidently results in an increase in the amount of entering fuel (path 1) to keep the production constant. Owing to the degradation, the new state of the outflow #2' is characterized by $(h_{2'}, s_{2'})$.

A differential analysis under the conditions: $W = \text{const.}$, $h_1 = \text{const.}$, and $s_1 = \text{const.}$, leads to the following expressions:

$$(h_1 - h_2)\, dm = m\, dh_2 \qquad [6.6]$$

$$(s_2 - s_1)\, dm + m\, ds_2 = dS_g \qquad [6.7]$$

Subsequently, if the *deterioration temperature* of flow #2 is defined as a quotient:

$$T_{d_2} \equiv dh_2/ds_2$$

the result is as follows:

$$\left(\frac{dm}{dS_g}\right)_r = \frac{T_{d_2}}{(h_1 - h_2) - T_{d_2}\,(s_1 - s_2)}, \qquad [6.8]$$

where r is the deterioration path, as shown in Figure 6.1.

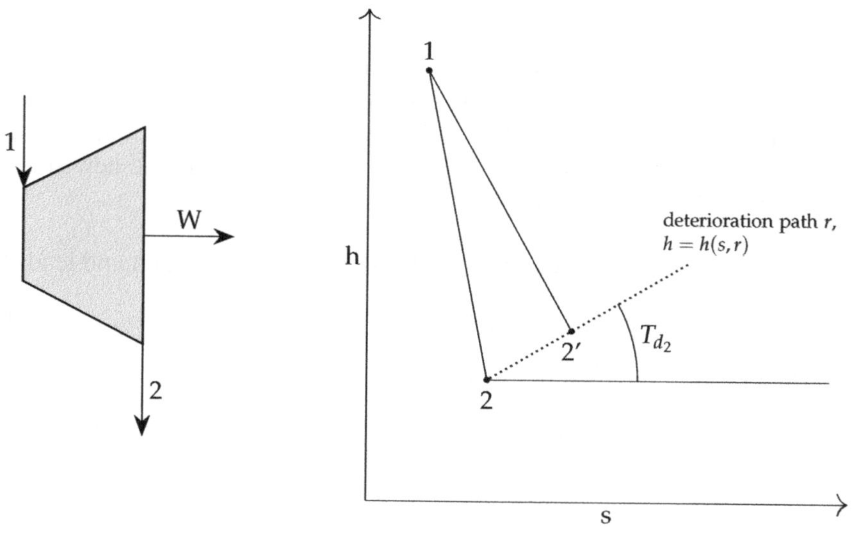

Figure 6.1. *The deterioration path of a turbine. For a color version of this figure, see www.iste.co.uk/feidt/advances.zip*

Equation [6.8] is the mathematical expression of the existing relationship between any local degradation and the increase in associated resources if the production of the component should remain constant.

These equations and the function $\ell \equiv (h_1 - h_2) - T_d(s_1 - s_2)$ were presented in Valero and Lozano (1992). Valero called it the *relative free energy* (Royo 1994; Royo and Valero 1995; Royo et al. 1997), and T_d the *deterioration temperature.* In 1994, Sciubba proposed to name it "dissipation" temperature. However, the authors consider the word "deterioration" more precisely, since that word always relates to a component rather than to a process.

The expression *relative free exergy* (RFE) originates from its similarity to the Gibbs free energy function. Equation [6.8] states that any deterioration process in an energy component has an associated T_d in the exit stream of supplying fuel. This parameter possesses temperature dimensions even if it is not a gauging temperature; it can be calculated by measuring the quotient dh_2/ds_2 experienced by the output stream of the component's fuel.

The heat exchanger (Figure 6.2) experiences the same procedure. In that case, the enthalpy and entropy balances are written as follows:

$$m_1 (h_1 - h_2) - m_3 (h_4 - h_3) = 0 \quad [6.9]$$

$$m_1 (s_1 - s_2) - m_3 (s_4 - s_3) = -S_g \quad [6.10]$$

If the heat source degrades, at equal to the specifications of the inflow #1, an increase in the entropy generation results evidently in an increase in the amount of entering fuel #1 to maintain a constant production; thus, m_3, (h_3, s_3) and (h_4, s_4) are constant. Owing to the degradation, the new state of outflow #2' is characterized by $(h_{2'}, s_{2'})$.

A differential analysis under the condition (h_1, s_1) is constant and leads to:

$$(h_1 - h_2)\, dm_1 = m_1 dh_2 \quad [6.11]$$

$$dS_g + (s_1 - s_2)\, dm_1 = m_1 ds_2 \quad [6.12]$$

If the deterioration temperature of flow #2 is defined as a quotient:

$$T_{d_2} \equiv \frac{dh_2}{ds_2} > 0$$

subsequently equation [6.8] is also satisfied.

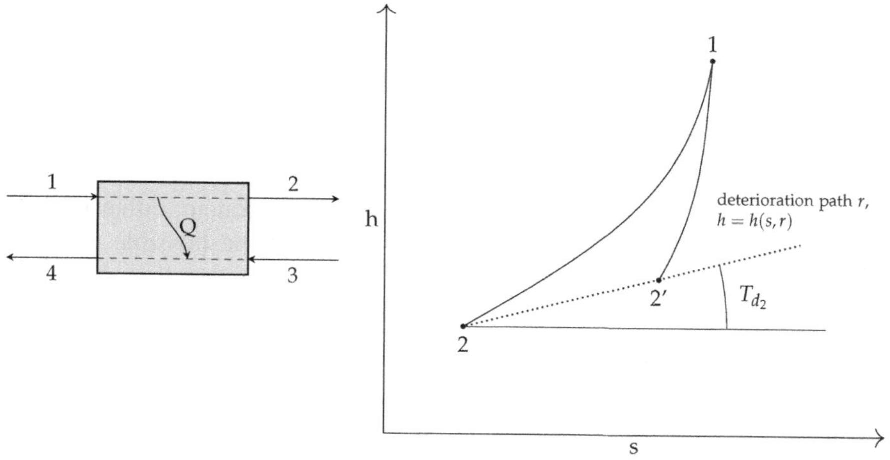

Figure 6.2. *Deterioration path of heat source in a heat exchanger. For a color version of this figure, see www.iste.co.uk/feidt/advances.zip*

According to the second law, dS_g is always positive; therefore, more fuel must be used to obtain the same product, and $dm > 0$. Because $T_{d,2}$ is positive, the term $(h_1 - h_2) - T_{d,2}(s_1 - s_2)$ must be positive; in other words, $T_{d,2}\,(s_1 - s_2)$ is always lower than $h_1 - h_2$. Also note that in a low pressure steam turbine of a Rankine cycle, $T_{d,2}$ coincides with the condenser temperature, which is fully logical. More examples could be found in Royo and Valero (1995).

6.4. h-s deterioration(s) path(s) of an energy system

Man-made energy systems transform the energy flows into other flows with a purpose. Commonly, the name of a component describes its function: a boiler boils water, a condenser condenses steam and a heat exchanger exchanges heat. The understanding of how nature behaves is the basis of the component design. The entering flow performs a purposive path through which its thermodynamic properties change progressively until the flow exits the component. This path is never strictly determined, either because of possible changes in the quantity/quality of the entering flow or because of changes in the behavior of the component itself. This path variation results in a change in the properties of the exiting flow with respect to its design conditions. There are as many new exit states as causes of change, which are called *malfunctions* and are classified into two categories: intrinsic and induced (Valero et al. 2004b; Valero et al. 1999; Torres et al. 1999). The term *intrinsic* refers to the internal deterioration of the component; the term *induced* refers to its off-design conditions (not the optimal operating conditions). A good component design foresees most of these off-design conditions. Literature and some manufacturers provide governing equations and control parameters based on semi-empirical models of the machines.

For a given deterioration cause r of an energy component, either intrinsic or induced, a geometric path in the $h - s$ plane of the possible exit flow states can be identified. Let $h_{2'} = h_{2'}(s_{2'}, r)$ to be the function describing this dissipation path of the exiting flow, as shown in Figure 6.1, where:

– $1 \longrightarrow 2$: is the design path of the stream in the component;

– $1 \longrightarrow 2'$: is the stream path after the component deterioration;

– $2 \longrightarrow 2'$: is the dissipation path of the exiting stream $h_{2'} = h_{2'}(s_{2'}, r)$.

The function $h_{2'} = h_{2'}(s_{2'}, r)$ can be regarded as the mathematical description of the effects on the exiting flow caused by the component deterioration cause r. This function always exists for a given differential malfunction interval. Moreover, under actual machine conditions, this function is a composite curve of the different paths caused by the concurrent deteriorations acting simultaneously on the component (Valero et al. 2004b). The next section focuses on this function.

6.5. The Legendre transform of a deterioration path

For each deterioration cause, r, the existence of a function $h = h(s)$ of the exiting flow can be assumed, which can be described in the $h - s$ plane with the conventional point geometry or Plücker line geometry for strictly convex functions as a convolution of the tangents of the original curve. This technique is named the *Legendre transform* of the function $h = h(s)$ and is widely used in classical thermodynamics (Callen 1985; Alberty 2001). As it is shown in Figure 6.3, the intersection of the tangent line of the deterioration path r with the h- axis is ℓ and its slope is T_d:

$$T_d = \frac{h - \ell}{s}$$

or

$$\ell = h - T_d\, s \qquad \text{i.e.} \qquad \ell = \ell(T_d) \tag{6.13}$$

Equation [6.13] is the *Legendre transform* of the deterioration path $h_{2'} = h_{2'}(s_{2'}, r)$. In fact, pairs of $(\ell_{2'}, T_{d,2'})$ for each exit state #2' of the component provide the same information as pairs $(h_{2'}, s_{2'})$. Since the relationship $\ell_{2'} = \ell_{2'}(T_{d,2'})$ is mathematically equivalent to the relationship $h_{2'} = h_{2'}(s_{2'}, r)$, it can also be considered the description of the deterioration path. In general, $h_r = h_r(s_r)$ is the deterioration function in the *h-representation*, whereas $\ell_r = \ell_r(T_d, r)$ is the deterioration function in the ℓ-representation. In other words, ℓ_r and $T_{d,r}$ are inseparable in a deterioration path r such as h_r and s_r do. Note also that h and s do not represent absolute values because they refer h_0 and s_0 values. Unless otherwise stated h and s represent $h - h_0$ and $s - s_0$.

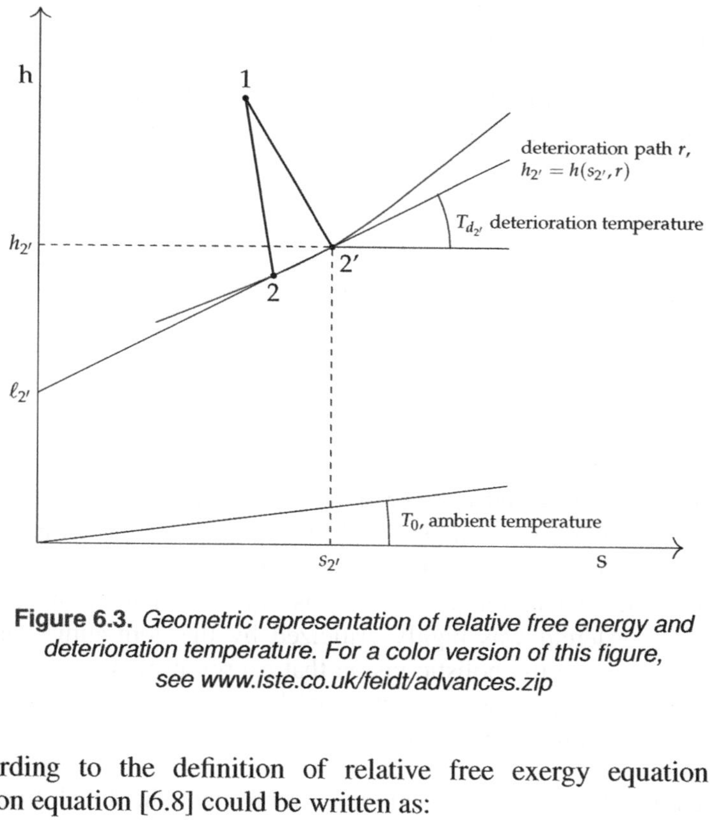

Figure 6.3. *Geometric representation of relative free energy and deterioration temperature. For a color version of this figure, see www.iste.co.uk/feidt/advances.zip*

According to the definition of relative free exergy equation [6.13], expression equation [6.8] could be written as:

$$\left(\frac{dm}{dS_g}\right)_r = \frac{T_d}{\ell_1 - \ell_2} \qquad [6.14]$$

The deterioration temperature T_d is not a physical temperature; it is a parameter and derivative with temperature dimensions:

$$T_d = \left(\frac{dh}{ds}\right)_r \qquad [6.15]$$

It can be interpreted as the entropic cost of the deterioration. The greater the T_d, the lower the relative free energy of the exiting flow. Well-known contributors to technical thermodynamics such as Baehr (2005) have defined the *friction work* or *loss work* for an expander as follows:

$$dW_r = T_d \, dS_g \qquad [6.16]$$

This loss work is energy that has not been dissipated into the environment at T_0 and contributes to the degradation of the exiting flow of the expander. It increases the enthalpy and entropy of the exiting stream and reduces the work produced by the turbine, thereby decreasing its efficiency.

In the example of the turbine (Figure 6.1), a differential analysis of the exergy equation [6.1] under the conditions (h_1, s_1) and W constants results in:

$$dF = dI = T_0 \, dS_g \qquad [6.17]$$

Therefore, the following expression is obtained:

$$dF = T_0 \, dS_g = \frac{T_0}{T_d} \, dW_r \qquad [6.18]$$

Another important antecedent was the work of Alefeld (1988), who worked on heat pumps and refrigeration systems: he questioned the use of the reference ambient temperature T_0 instead of a variable temperature T_x associated with the whole system. In particular, he proposed for the case of a Rankine cycle, where temperature T_x is equal to that of the condenser vacuum, T_{vac}. Alefeld was highly criticized by the community of exergy practitioners, because the holistic vision that exergy provides was lost in his proposal. This chapter is a recognition of his work because mathematical evidence is stronger than reasonable assumptions.

The relative free energy function, ℓ, represents the local free energy conditioned by the deterioration path r. It is an intensive property, and its corresponding extensive property:

$$\mathcal{L} = m \, \ell \qquad [6.19]$$

is the amount of energy of the exiting stream in a fuel of the type $E_1 - E_2$, which is not affected by the deterioration of the component. This is because when the exit flow changes its state owing to the deterioration, h and s, vary by following the path $h_r = h_r(s_r)$. Consequently, the following expression is obtained:

$$d\ell_2 = dh_r - \left(\frac{dh}{ds}\right)_r ds_r = 0 \qquad \text{or} \qquad \ell_2 = \text{const.} \qquad [6.20]$$

In addition, the function $h_r = h_r(s_r)$ is neither a straight curve nor a slope T_0. Therefore, ℓ exceptionally coincides with the specific exergy of the stream, and the exergy increase in the output stream in the turbine is:

$$ex_{2'} - ex_2 = (h_{2'} - h_2) - T_0\,(s_{2'} - s_2).$$

In the example of the turbine, (h_1, s_1) and W are constant, then applying differential analysis to the unit exergy consumption definition: $k \equiv F/P$, it follows:

$$dF = P\,dk \qquad [6.21]$$

Combining both equations [6.17] and [6.21]:

$$dk = \frac{T_0\,dS_g}{W} = \frac{T_0}{T_d} \cdot \frac{T_d\,dS_g}{W} \qquad [6.22]$$

and applying equation [6.14], it follows:

$$\left(\frac{dk}{dm}\right)_r = \frac{T_0}{T_d} \cdot \frac{\ell_1 - \ell_2}{W} \qquad [6.23]$$

This expression is quite important, because it relates the loss of exergy efficiency in the turbine with the additional mass of fuel to compensate it at constant production. In other words, it relates the malfunction of the turbine with the need for more entering fuel. Also, this expression opens the way of changing the paradigm of exergy by relative free energy.

Therefore, a new definition of the efficiency (or its inverse specific consumption) of the turbine appears:

$$k_L = \frac{F_L}{P_L} = \frac{\mathcal{L}_1 - \mathcal{L}_2}{W} = \frac{H_1 - H_2 - T_d\,(S_1 - S_2)}{W} \qquad [6.24]$$

This new definition of efficiency of the turbine has the important property that if the turbine malfunctions following a r path, its characteristic equation becomes:

$$\mathcal{L}_1 = \mathcal{L}_2 + k_L\,W + \text{const.} \qquad [6.25]$$

this constant might be selected as the standard reference state for h and s calculations (h_0, s_0).

When a malfunction in the turbine follows a trajectory r at W constant, it follows:

$$\ell_1\, dm = \ell_2\, dm + W\, dk_L \quad [6.26]$$

because, $\ell_1 - \ell_2$ is constant along trajectory r. Therefore:

$$\frac{dk_L}{k_L} = \frac{dm}{m} \quad [6.27]$$

in other words, any additional percentage of malfunction following a trajectory r at W constant result in an additional percentage of mass of fuel entering the turbine.

Also note that exergy does not fulfill equation [6.27] because:

$$ex_1\, dm = ex_2\, dm + m\, dex_2 + W\, dk \qquad dex_2 \neq 0$$

Equation [6.25] is not a mere efficiency definition, as it is the unit exergy consumption definition $k \equiv F/P$. It corresponds to a physical description of the turbine system and what is more important, it does not depend on the reference environment temperature T_0, but on the behavior of the turbine. The structural theory, together with the relative free energy function, allows to obtain an interesting characteristic function of the behavior of thermodynamic systems. If under the structural theory, the coefficients k_{ij} are the precursors of costs k_j^*, should these costs be their *natural costs* or perhaps they should be named *regular costs*. In any case, these costs are directly related to the additional amount of resources needed to compensate for the malfunction.

The expression of equation [6.14] shows that the second law does not link the relationship between thermodynamic quantity, quality, cost, and irreversibility. In general, the reason for using T_d instead of T_0 arises from the fact that the additional degraded energy not used in the production of the product becomes part of the output stream instead of going directly to the environment at T_0. The value of T_d depends on how large the additional irreversibility is. This interpretation of T_d helps to explain that its lowest achievable value is T_0, while it apparently has no upper limit.

In other words, the deterioration lowers the free energy of the exiting stream. As this deterioration always relates to some component, this is why we name this function relative free energy. Similar to the Gibbs free energy that relates to some chemical reaction, the RFE relates to some component deterioration in a system.

The key problem of diagnosis of energy systems is how to assess, as precisely as possible, the relationship between local irreversibilities and additional consumption of resources. At constant production, a local irreversibility needs additional resources to compensate it, but also modifies the exiting streams, affecting downstream enthalpies and entropies and eventually the upstream ones, in case of recycling. A certainty derives from the second law: Any inefficiency in a process increases the entropy generation. This means that any mechanical, thermal or chemical loss generates entropy and reduces efficiency. However, any efficiency decrease means more resources (at equal quality) to get the same product. If one knows the amount of local resources needed to compensate a local deterioration (r), that is:

$$\left(\frac{\partial m_{local}}{\partial S_g}\right)_r$$

and knows how the global system's resources are related to local resources, that is:

$$\left(\frac{\partial m_{global}}{\partial m_{local}}\right)_r$$

one may have a new theory of energy system diagnosis.

6.6. Conclusion

The definition of efficiency plays a fundamental role in the process of calculating average exergy costs. So, efficiency and cost became closely linked. In such a way, the *theory of exergy cost* (Valero et al. 1986; Lozano and Valero 1993) was extensively used to assess costs on many thermoeconomic/thermodynamic applications. However, the exergy cost depends on the plant disaggregation scheme. It is necessary to look after the cost formation process by searching how much exergy has been consumed in the processes and where they have consumed it. The result of this activity is a given productive structure. However, if exergy depends on a chosen reference state, exergy efficiencies and their closely related costs will depend on such selection. Notwithstanding, any deterioration that increases entropy generation results in an additional expense of resources used to produce the same product. Both the entropy increase and the additional resources are measurable and free from analyst selections. Therefore, something fails in the conventional theory.

The use of exergy instead of enthalpy or the Gibbs function for obtaining costs increase the overall, coherent and systematic view of a set of stream costs in a productive structure. To what extent could we use these costs to assess component malfunctions? In other words, can average costs be used as marginal costs?

The exergy cost is the exergy backpack, embodied exergy or the exergy footprint of a commodity; in fact, Szargut named it as *cumulative exergy consumption* (Szargut et al. 1988), that is, its past characterization (history), while exergy is the potential energy of such a commodity, or its remaining capacity for doing something else. Can average costs predict future degradations or only past irreversibilities?

The purpose of thermoeconomics is to obtain a coherent and significant set of costs in a given structure. That is because its applications are directed toward the optimization and diagnosis of productive structures. Therefore, the costs of internal streams are quite important to analyze the behavior of the system. Consequently, internal costs ought to be as sensitive as possible to system degradation. A theory that would provide exact costs would be the objective of thermoeconomic diagnosis. In fact, one can obtain the costs of final products or commodities, without performing very detailed analyses. In other words, in no way thermoeconomics, as it is already known, is far from being finished. Moreover, the analyst needs to look for the *cost formation process of wastes* (Torres et al. 2008), not mentioned here even when some ideas developed here could be used to clarify it.

These two chapters present a general theory of (linear) costing, demonstrating and answering questions such as:

– To what extent of the linear characteristic equation, can a component be conformed to its efficiency definition, $F - k\,P = 0$?

In fact, exergy efficiency must be coherent with the component's design purpose, but we design machines by observing the behavior of nature. So what is first, efficiency or nature? We overcome this drawback with equation [6.24].

– Why the F and P rules or any other costing proposal may be either rational or not under a given disaggregation scheme?

– Under a set of conditions, we may say that the exergy costs balances are the mathematical dual of exergy balances and vice versa. The exergy cost and the exergy are like specular images of the same entity.

– A costing theory applicable to any thermodynamic function such as enthalpy, exergy, Gibbs free energy and so on.

– In analyzing this last point, we proposed a new thermodynamic function, called the *relative free energy*, ℓ, and introduced a new parameter T_d called deterioration temperature due to a component's deterioration cause r, characterized by a thermodynamic trajectory $h = h(s,r)$ describing the effects on the exiting stream.

– The Legendre transform of the deterioration path, $h_2 = h_2(s_2,r)$ is the relationship $\ell_2 = \ell_2(T_{d,2},r)$. In other words, the pairs ℓ_r and $T_{d,r}$ are so inseparable in a deterioration path r, as the pairs (h_r, s_r) do.

– The general formula [6.14] applies for different flowing streams in energy systems under each particular deterioration path, r_i. A given component may have several deterioration paths, not only one. Therefore, the concept of component malfunction need be associated with that of deterioration paths. Besides that, this formula can be applied to any system – under the defined conditions – no matter its aggregation level. Each degradation will have its deterioration temperature as well as its corresponding RFE efficiency.

– A way to assess appropriate characteristic equations of thermal systems as it was obtained here for the case of turbines.

– As the component's deterioration does not depend on the analyst's decision on the chosen reference state, the deterioration temperature no longer depends on the chosen reference temperature. Neither does the new definition of efficiency using the RFE. This fact does not isolate the plant from environmental conditions, but eliminates the arbitrary selection of any reference state. That idea clarifies the role of the environmental conditions onto the plant.

The exergy cost theory provides reasonable costs. Will relative free energy provide a set of natural (or perhaps named regular) costs? In fact, the cost of a deterioration is directly related to the physical behavior of that deterioration. This way it also means that these costs are in fact marginal costs, since they incorporate the future malfunction behavior of the component.

The theory sketched here opens new fields of knowledge, since new questions appear: is exergy the best thermodynamic function when diagnosing systems? Why does T_0 need to be the same for each component of a given structure? Can we use this theory to assess objective average costs of deteriorations free from assumptions? The deterioration behavior of an energy system relates with internal marginal costs while exergy costs with history. Should we change our definitions of efficiency under the light of the relative free energy function? etc. Even if, as described early in the 1990s, all these ideas are tools for future (Valero 2000). Cost is a measure of expended resources to produce something, then, costing with the *relative free energy* instead of exergy would open a new field of a more precise theory of thermoeconomics. We presently work with such ideas.

6.7. Epilogue

The creation of a new thermodynamic function is not a common event. All thermodynamics books consider the internal behavior of systems when they are exposed to external forces. Thus, their measurable properties such as pressure, temperature, volume, composition, concentration and others change, affecting their energy and entropic quantities. According to the formulation of Callen and followers (Callen 1985), there is always a fundamental equation, $U = U(S, V)$, which links the properties of internal energy, entropy and volume of a system in all its equilibrium states. From there, if it were known, we could obtain the rest of the properties by derivation and by Legendre transforms, obtaining, in addition, Maxwell's equations.

Thus, conventionally, we obtain the well-known expressions for a single-component system:

$$T = \left(\frac{\partial U}{\partial S}\right)_V \qquad [6.28]$$

$$P = -\left(\frac{\partial U}{\partial V}\right)_S \qquad [6.29]$$

and

$$H = U + PV \qquad [6.30]$$

$$F = U - TS \qquad [6.31]$$

$$G = H - TS \qquad [6.32]$$

Or in its differential form:

$$dU = T\,dS - P\,dV \quad [6.33]$$

$$dH = T\,dS + V\,dP \quad [6.34]$$

$$dF = P\,dV - S\,dT \quad [6.35]$$

$$dG = V\,dP - S\,dT \quad [6.36]$$

This elegant formulation is able to condense the equilibrium thermodynamic properties of any system. However, in fundamental thermodynamics, the system is always the object of study rather than the system's interaction with the machine to perform a designed path. One step along this way is to consider the environment. This reservoir of energy attracts systems that are in disequilibrium with it. Therefore, a new branch of thermodynamics has been developed, which studies the behavior and properties of systems in relation to the surrounding environment. For this purpose, the concept of exergy was introduced:

$$B = (H - H_0) - T_0(S - S_0) \quad [6.37]$$

Now the properties of the system are confronted with those of the environment (T_0, H_0, S_0). It no longer matters how much energy a system has but how much energy it has with respect to the environment. Thus, we call a resource a matter or energy flow whose thermodynamic properties differ from those of the environment. Waste is even a resource because it has exergy that can potentially harm the environment. So much so that, in an ecological view of thermodynamics, the object of interest is the environment and how to preserve it from the waste we release from our physicochemical processes.

Indeed, we design physicochemical processes to transform resources into products, and the equipment we use interacts with the working substances, not only under design conditions but also when they are forced to work outside those conditions and/or degrade.

We know from economics that if some equipment degrades or malfunctions, it will have an added cost, conventionally measured in monetary units. However, money is a poor evaluator of degradation because it can be created by human will, whereas energy can neither be created nor destroyed by human agency. Resources have exergy and what is expended is exergy that can be measured in kWh, independent of money. It allowed the creation of the derived concept of exergy cost (Valero et al. 1986), and unconnectedly, Szargut called *cumulative exergy consumption* (Szargut et al. 1988) of a product/system, which is no longer only dependent on the environment, but also on the previous history of its formation. The exergy cost is the addition of all the exergy resources that have been used to produce it. It is an emergent property, that is, it is a property that cannot be measured from the current properties of the system, but it is necessary to describe the processes that the product has undergone to become what it is. However, with exergy, it was necessary to consider the properties of the environment, with the exergy cost, it is also necessary to know the processes of the formation of a product, measured in physical units.

It is clear that the rigor of equilibrium thermodynamics has been lost because the analyses are no longer objective. They depend on the analyst, fortunately other problems unknown to science became solved.

The new properties, exergy and exergy cost, make it possible to tackle, through thermoeconomics, the problem of diagnosis. In other words, it is a question of analyzing how many additional resources are consumed and in which equipment they are consumed when an industrial energy system works outside its design conditions. The industrial system, composed of a set of apparatuses, transforms resources into products and waste using physicochemical processes.

Costing the causes of deterioration needs new insights, which malfunction and dysfunction are (Valero et al. 2004b). These terms allow us to analyze the causality of the degradation(s) of a component(s) and the impacts assumed by the rest of the system, measured in energy units. We call this malfunction the deterioration of a component measured in terms of exergy loss or irreversibility. At the same time, dysfunction measures the increase in irreversibility that each and every other component has to endure to adapt to the new situation. Although much work has been done on this problem using exergy and its conceptual derivatives, exergy has an inherent drawback. When calculating exergy, one chooses a reference state that depends on the analyst, not the machine. But the deterioration of any component is a physical effect intrinsic to the machine, not to the analyst. It is an interaction between the machine that degrades and the working fluid that makes it work, the result of which is the use of more local resources to produce the same thing. In other words, it costs more locally to produce the same thing. However, this localized additional cost or consumption in the component must be passed on to the system as a whole. Therefore, we will continue to use the concepts of malfunction and dysfunction, but their quantification must be re-examined.

Here, the concept of *relative free energy* arises as a new thermodynamic function. The Gibbs free energy, $H - T\,S$, can be interpreted as the free energy of an open system resulting of subtracting from its enthalpy a quantity depending on its own temperature multiplied by its disorder measured in entropic units.

The higher the temperature of a system, the lower the amount of absolute energy H will be converted into useful energy produced by the system. We observe that nothing is said about the machine which can convert it. The machine is here an external element. However, when T is replaced by T_0, and in addition, the energy H and entropy S of the system are subtracted from their respective amounts when the system reaches an equilibrium with the environmental attractor, that is, by H_0 and S_0, the amount of available energy is limited. Again, exergy is free energy, that is, available energy can be converted for other uses. Here, we are not yet talking about the machine that will convert it, but about the potential of the system in relation to the environment. In other words, exergy is a measure of the thermodynamic distinction of the system from its environment. It is already the result of the interaction between the system and its environment. It is, in fact, a free energy *relative* to the environment.

Industrial systems are composed of equipment assembled together to produce one or more products. Moreover, in each component, there is a local interaction between the working fluid and its machinery. If each subsystem has its own *environment*, why reduce it to one? The local and available energy is limited by the machine that processes the fluid, not by the global environment. It is the reason for including the new function we call relative free energy. Then one uses T_d, instead T_0 or T, to limit the energy.

In this chapter, we have seen some properties of the RFE, and we have also shown how it is related to the deterioration temperature through the Legendre transform, in the same way as the intrinsic functions of systems in equilibrium such as the internal energy, the enthalpy, the Helmholtz function and the Gibbs function. It is a significant conceptual advancement but still needs development. It was first presented in 1992 at the ECOS conference (Royo and Valero 1995), with some preliminary work by Alefeld at the University of Munich (Alefeld 1988) as indicated in the introduction of this chapter. However, the use of this function in diagnostic applications and cost assessment of malfunctions in industrial energy systems are still to be developed. We hope that this work will encourage researchers to write other chapters on thermodynamics and thermoeconomics.

6.8. References

Alberty, R. (2001). Use of Legendre transforms in chemical thermodynamics: (IUPAC technical report). *Pure and Applied Chemistry*, 73(8), 1349–1380.

Alefeld, G. (1988). Problems with the exergy concept (or the missing Second Law). *IEA Heat Pump Newsletter*, 6(3), 19–23.

Baehr, H.D. (2005). *Thermodynamik*. Springer, Berlin, Germany.

Callen, H.B. (1985). *Thermodynamics and an Introduction to Thermostatistics*, 2nd edition. Wiley, New York, USA.

El-Sayed, Y.M. and Evans, R.B. (1970). Thermoeconomics and design of heat systems. *Journal of Engineering for Power*, 92, 27–35.

Georsgecu-Roegen, N. (1971). *The Entropy Law and the Economic Process*. Harvard University Press, New York, USA.

Javier, R. (1994). Las ecuaciones caracteristicas de los sistemas térmicos. La energía libre relativa. PhD Thesis, Universidad de Zaragoza, Spain.

Juan Martinez-Alier, K.S. (1987). *Ecological Economics: Energy Environment and Society*. Basil Blackwell Ltd, New York, USA.

Lozano, M. and Valero, A. (1993). Theory of the exergetic cost. *Energy*, 18(9), 939–960.

Royo, J. and Valero, A. (1995). Towards a unified description of the energy behavior of thermomechanical systems. In *Analysis and Improvement of Energy Systems*, Krane, R.J. (ed.). ASME, New York, USA.

Royo, J., Valero, A., Zaleta, A. (1997). The dissipation temperature: A tool for the analysis of malfunctions in thermomechanical systems. *Energy Conversion and Management*, 38(15–17), 1557–1566.

Sciubba, E. and Wall, G. (2007). A brief commented history of exergy from the beginnings to 2004. *International Journal of Thermodynamics*, 10(1), 1–26.

Szargut, J., Morris, D.R., Steward, F. (1988). *Exergy Analysis of Thermal, Chemical, and Metallurgical Processes*. Hemisphere, New York, USA.

Torres, C., Valero, A., Serra, A., Royo, J. (1999). Structural theory and thermoeconomic diagnosis. Part I: On malfunction and dysfunction analysis. In *Proceedings of the ECOS'99, The American Society of Mechanical Enegineers*, Tokyo, Japan.

Torres, C., Valero, A., Rangel, V., Zaleta, A. (2008). On the cost formation process of the residues. *Energy*, 33(2), 144–152.

Tribus, M. and Evans, E. (1962). A contribution to the theory of thermoeconomics. Technical Report 62, University of California, Los Angeles, CA, USA.

Tsatsaronis, G. and Winhold, M. (1985). Exergoeconomic analysis and evaluation of energy-conversion plants I: A new general methodology. *Energy*, 10(1), 69–80.

Valero, A. (2000). El marco termodínamico para iluminar la sociedad actual. In *Economía, ecología y sostenibilidad en la sociedad actual*, Naredo, J.M., Parra, F. (eds). Siglo XXI de España Editores, Tres Cantos, Spain.

Valero, A. (2018). The duality between exergy efficiencies and exergy costs. The structural theory and the relative free energy function. In *Proceedings of the Journées sur le thème de l'éxergie*, Nancy, France, 22–23 November.

Valero, A. and Lozano, M. (1992). General theory of thermoeconomics. Part II: The relative free energy function. In *Proceedings of ECOS '92 International Symposium*, Zaragoza, Spain.

Valero, A. and Torres, C. (2020). Relative free energy function and structural theory of thermoeconomics. *Energies*, 13(8), 2024.

Valero, A., Lozano, M., Muñoz, M. (1986). A general theory of exergy saving. Part I: On the exergetic cost. *Computer-aided Engineering and Energy Systems: Second Law Analysis and Modelling*, 3, 1–8.

Valero, A., Torres, C., Serra, L. (1992a). General theory of thermoeconomics. Part I: Structural analysis. In *Proceedings of ECOS '92 International Symposium*, Zaragoza, Spain.

Valero, A., Torres, C., Serra, L. (1992b). General theory of thermoeconomics: Part II. The relative free energy function. In *Proceedings of ECOS '92 International Symposium*, Zaragoza, Spain.

Valero, A., Torres, C., Lerch, F. (1999). Structural theory and thermoeconomic diagnosis. Part III: Intrinsic and induced malfunctions. In *Proceedings of the ECOS '99*, Tokyo, Japan.

Valero, A., Correas, L., Lazzaretto, A., Rangel, V., Reini, M., Taccani, R., Toffolo, A., Verda, V., Zaleta, A. (2004a). Thermoeconomic philosophy applied to operating analysis and diagnosis of energy utilty systems. *Int. J. Thermodynamics*, 7(2), 33–39.

Valero, A., Correas, L., Zaleta, A., Lazzaretto, A., Verda, V., Reini, M., Rangel, V. (2004b). On the thermoeconomic approach to the diagnosis of energy system malfunctions. Part 2: Malfunction definitions and assessment. *Energy*, 29(12–15), 1889–1907.

List of Authors

Monica COSTEA
Department of Engineering
Thermodynamics,
Heat Engines, Thermal and
Refrigeration Equipment
Politehnica University of Bucharest
Romania

Michel FEIDT
LEMTA
Université de Lorraine
Vandoeuvre-les-Nancy
France

Stoian PETRESCU
Department of Engineering
Thermodynamics,
Heat Engines, Thermal and
Refrigeration Equipment
Politehnica University of Bucharest
Romania

Julien RAMOUSSE
LOCIE
UMR 5271
CNRS
Université Savoie Mont Blanc
Bourget du Lac
France

César TORRES
CIRCE Institute
University of Zaragoza
Spain

Antonio VALERO-CAPILLA
CIRCE Institute
University of Zaragoza
Spain

Index

L, O, P

R, S, T

Printed and bound by CPI Group (UK) Ltd, Croydon, CR0 4YY

13/09/2026

14960999-0001